2018年度山东省本科教改项目（编号：Z2018S002；重点项目）：
基于CBI理念的大学英语教学新模式与综合评价体系建设研究

高等学校“十三五”规划教材
服装设计与工程国家级特色专业建设教材

服装专业英语

主　编　王桂祥
副主编　马应心

北　京
冶金工业出版社
2019

内容提要

本书系统介绍了服装的设计、生产、面料、演示以及流行趋势预测等基本知识。在结构上，本书配合常规的课时安排，共设置11个单元，方便教师课堂授课。每单元主文为师生提供了一些必要的专业常识、新兴资讯以及常用专业用语，使学生既能掌握基本的服装英语，又能对服装业的概况有基本的认识。

本书为服装设计与工程国家级特色专业建设教材，并可供相关专业的从业人员阅读参考。

图书在版编目(CIP)数据

服装专业英语/王桂祥主编.—北京：冶金工业出版社，2019.9

高等学校"十三五"规划教材

ISBN 978-7-5024-8224-4

Ⅰ.①服… Ⅱ.①王… Ⅲ.①服装工业—英语—高等学校—教材 Ⅳ.①TS941

中国版本图书馆CIP数据核字(2019)第186747号

出 版 人　谭学余

地　　址　北京市东城区嵩祝院北巷39号　邮编　100009　电话　(010)64027926

网　　址　www.cnmip.com.cn　电子信箱　yjcbs@cnmip.com.cn

责任编辑　杜婷婷　美术编辑　郑小利　版式设计　孙跃红

责任校对　王永欣　责任印制　牛晓波

ISBN 978-7-5024-8224-4

冶金工业出版社出版发行；各地新华书店经销；三河市双峰印刷装订有限公司印刷

2019年9月第1版，2019年9月第1次印刷

787mm×1092mm　1/16；10.25印张；244千字；154页

35.00元

冶金工业出版社　投稿电话　(010)64027932　投稿信箱　tougao@cnmip.com.cn

冶金工业出版社营销中心　电话　(010)64044283　传真　(010)64027893

冶金工业出版社天猫旗舰店　yjgycbs.tmall.com

(本书如有印装质量问题，本社营销中心负责退换)

本书编委会

序 言

纵观服装产业，近几年一直处于“朝阳”状态。我国既是全世界最大的服装生产国之一，也是全世界最大的服装生产加工基地之一。根据智研咨询网发布的《2017—2023年中国服装零售产业深度调研及未来发展趋势报告》：受益于世界经济的持续复苏和海外消费逐渐回暖，中国服装行业生产、出口数据整体向好。参考国外服装巨头体量，我国单个服装企业仍存在发展壮大的空间。

在国内服装产业集群发展趋向品牌化、规模化、国际化的产业转型升级大背景下，既熟悉服装生产与经营，又熟练掌握服装专业英语的人才，成了服装企业急需的“香饽饽”。

为了适应目前的经济情况，为服装行业培养更多的综合性人才，本书作者根据自己多年的教学实践经验，精心编写了这本教材。同时，随着我国教育教学改革的不断深入和科学技术的飞速发展，部分服装专业英语教材内容已经稍显陈旧，需要更新。基于此，本书介绍了最新的服装面料、设计、裁剪和色彩等内容。为方便读者在有限的时间内更为有效地学习实用的服装专业英语，本书采用原汁原味外文材料，用英文的形式（配参考译文）介绍了服装行业研发设计、面料加工以及时尚秀等内容，可即学即用，增强职业竞争力。

德州学院　**徐静**

2019年9月6日

前 言

本书作为高等学校服装设计与工程国家级特色专业建设教材，主要面向高等学校高年级服装设计与工程专业的学生。在服装专业英语教学活动中，我们发现了两个问题：一是服装专业英语课程理论性太强，涉及面过宽，教师对课程的讲授深度产生困惑；二是学生在具体的学习过程中对课程内容的理解有一定的困难。鉴于教师教学与学生学习这两方面的问题，我们编写了本书。

学习本书的目的在于培养学生的服装英语专业能力，切入点是“服装专业英语”课程理论教学中的难点，通过多种类型的题目训练，提高学生对服装专业英语的理解。本书内容丰富，用11个单元对知识点进行了解析，希望学生在服装专业的学习或实践中遇到疑惑时，通过学习本书知识点的解析能够获得一些帮助。

本书共分为11个单元，每单元内容分两部分，第一部分为本章学习材料；第二部分为练习题，包括填空题、简答题、翻译、口头汇报及写作5种题型。

本书在编写过程中得到了很多专家、教授的指点和帮助。德州学院副校长李永平教授对本书编写给予了中肯的建议，德州学院教务处处长、山东省纺织与轻工业专业教学指导会委员徐静教授在百忙之中对本书内容进行了修改并为本书作序，德州学院纺织服装学院马洪才院长审阅了全书，在此对他们表示衷心的感谢。

服装专业英语是一门发展中的课程，由于时间仓促和编者水平所限，书中不妥之处，敬请广大读者批评指正，并诚挚地欢迎读者提出宝贵建议。

编　者

2019年7月

Contents

Throughout history there have been associations between clothing and textiles. This relationship is not only simply one of being connected, in many ways they are inseparable. When considering one it is very difficult to ignore the other. This intimate bond between the two fields is revealed when changes occurring in one ultimately, if not immediately, have some impacts on the other. Perhaps now more than ever the relationship is at its closest, largely due to technological advances, great cha[illegible]es in lifestyle and the growing sophisticat[illegible] of consumer demands. The current ra[illegible]istic[1] influences reflected in tex[illegible]n goes hand in hand with adv[illegible]ated technology and scientific research. [illegible]nomenon mirrored perhaps in an ea[illegible] when Du Pont's development of [illegible]d to the craze for nylon stockings [illegible]e 1940s. It is evident that fabrics and fas[illegible]ons are marked with an indelible[2] seal [illegible]dicating their era and [3]mutual origins.

Unit 1 Textile

第1单元 面料

Throughout history there have been associations between clothing and textiles. This relationship is not only simply one of being connected, in many ways they are inseparable[1]. When considering one, it is very difficult to ignore the other. This intimate bond between the two fields is revealed when changes occurring in one ultimately, if not immediately, have some impacts on the other. Perhaps now more than ever, the relationship is at its closest, largely due to technological advances, great changes in lifestyle and the growing sophistication of consumer demands. The current rage for futuristic① influences reflected in textiles and fashion goes hand in hand with advances in related technology and scientific research. A phenomenon mirrored[2], perhaps, in an earlier time when Du Pont's development of Nylon led to the craze[3] for nylon[4] stockings[5] in the 1940s. It is evident that fabrics and fashions are marked with an indelible② seal indicating their era and mutual③ origins.

The relationship between fashion and textiles involves association between fashion designer, textile designer, fiber industries, color authorities, global commerce, the consumer and he media industries to mention but a few. As the raw material[6] of fashion, textiles constitute a complex system of primary industries comprising fiber and fabric producers[7] of both natural and artificial[8] materials and including the research, development and finishing industries. The fiber and textile industries have grown and expanded rapidly in recent years, facilitated by technological and mechanical advances in everything from the origination of new fibers through to innovative methods of production. These important and sometimes groundbreaking changes, in turn, continually provide the fashion industry with a vast assortment[9] of materials to choose from.

1.1 Fashion design and textiles

The textile industries collaborate[10] with a whole host of professionals, all experts within their own fields, from fashion to color and fibers. These experts play a pivotal role in the textile and fashion interrelationship④: they guide the textile companies, designers and technologists, advising them about future ranges and predicting why or how their ranges will appeal to the consumer.

Part of the mechanics of the relationship between fashion and textiles is rooted in trends and these influences the preliminary⑤ stages of fiber production (which generally operate two years ahead of a season). Trends in fibers and fabrics develop from information gathered from professionals in fashion, textile mills, and other industry experts. From this primary level, color and fabric trend information is then disseminated throughout the fashion and textile industries. Then initial judgments and choices in textiles focus on color. At this early stage views and considered opinions are drawn from industry experts such as the International Colour Authority (ICA) and the Color Association of the United States. The fiber and fabric industries also resolve issues of texture, production and construction, which are typically informed by demands[11] within fashion. The textile industry is steered by fabrication requirements that the fashion designer stipulates⑥. The fashion designer' relationship with fabric can be intensely⑦ personal. This intensity is very apparent at haute couture level, more so than at any other level, and is largely due to the fact that indulgence and personal expression can be afforded at this level.

服装和面料的关系贯穿整个历史。它们之间的关系不仅仅是彼此关联，在很多方面密不可分，考虑一方面，很难忽略另一方面。两者之间的亲密关系表现为其中之一发生变化，就会对另一个产生影响，即使影响不是立即产生，但最终会有影响。当今两者的紧密程度前所未有，很大程度上由于技术进步、生活方式的巨大变化和消费者需求的日益成熟。在未来主义潮流的影响下，面料和时尚在技术和科学研究上齐头并进地发展。早期的一种现象就是例证，20 世纪 40 年代杜邦公司开发的尼龙掀起了尼龙长袜的狂热。很明显，面料和服装都有不可磨灭的印记表明其时代和共同的起源。

时尚和面料的关系涉及时尚设计师、面料设计师和纱线行业，关联色彩全球贸易、消费者和媒体行业，等等，不胜枚举。作为时尚的原材料，面料的生产构成了原始材料公司的复杂体系，包括天然和人工材料的纱线和面料生产商，又进一步分为研究、开发和后处理公司。在最近几年，纱线和面料公司都得到了迅速的成长和发展，这归因于先进的技术和设备，促进了从新型纤维的开发到产品的创新方法等诸多方面的进步。这些重要的、有时是突破性的变化，反过来，不断地为时尚行业提供更多可选择的材料。

1.1 时尚设计和面料

面料行业与一系列专业人士通力合作，他们是时尚、色彩、纤维等自身领域的专家。这些专家在面料和时尚的相互关系中发挥了举足轻重的作用：他们指导面料公司、设计师和技术人员，为他们未来产品的范围提出建议，预测他们的产品为什么或如何吸引消费者。

从技术的角度看，时尚和面料的关系可以追溯到源头，即纤维的流行趋势和纤维产品的影响（通常在时尚季节前两年开始生产）。纤维和面料的流行趋势集中时尚专家、面料公司和其他专业公司的信息而发展。色彩和面料的流行趋势信息从源头层面上传播到时尚和纺织公司。然后，纺织公司对色彩作初始的判断和选择。在早期阶段，观点和参考意见通常来自专业机构，例如国际色彩权威和美国色彩协会。纤维和面料公司还要解决肌理、生产和构造问题，通常从时尚中了解需求。时尚设计师对面料的具体要求起着导向纺织公司的作用。时尚设计师与面料有着非常个性化的关系，在高级时尚层面上尤为密切，超过其他任何层面，很大程度上得益于那些买得起的人沉溺个性表达。

专业词汇

1. inseparable n. 不可分的
2. mirror vt. 折射
3. craze vt. 发狂
4. nylon n. 尼龙
5. stocking n. 长筒袜
6. raw material 原料
7. producer n. 生产商
8. artificial adj. 人造的
9. assortment n. 分类
10. collaborate vi. 合作
11. demand n. 需求

通用词汇

① futuristic adj. 未来主义的
② indelible adj. 擦不掉的
③ mutual adj. 相互的
④ interrelation n. 相互关系
⑤ preliminary adj. 初步的
⑥ stipulate vt. 规定
⑦ intense adj. 热情的

The size of a fashion company, and the type of designer within that company, dictates the attitude they have toward fabrics. For example, a company with a large market share will exercise a safe policy of "repeat" ordering[12]. Successful fabrics from previous seasons are ordered at a very early stage in the design process. Operating this way, the designers in these larger companies are familiar with the performance characteristics of chosen fabrics and are able to avoid the draping part of the design process. Draping being a normal way to find out how the material behaves on the body. This is in stark contrast to the smaller fashion design company that offers its clientele something original and unique[13]. As a consequence of using a greater variety of fabrics and less repeat[14] fabrics, they need to assess the draping quality of fabrics more than large companies. The designer within a large company is able to sustain[15] good relations with fabric producers facilitating communication about characteristics sought or the nature of emergent trends, this can result in the manufacture of new fabrics in order to update[16] the company's ranges. In contrast, smaller companies tend to focus on the individuality their products imply, achieving this through fabric selection as well as silhouette. The fashion designer's fabric selections are relative to the market they are supplying and they will ultimately choose the materials they judge to be most desired and appropriate for their consumer.

For the larger company, cost[17] factors are crucial when operating to tight profit margins[18], therefore, these companies will seek cost effective solutions[19] as regards their designing and fabric sourcing. Necessarily larger companies will also benefit from economies of scale. These are key factors if they wish to sustain market share[20] and healthy profits. At the other end of the spectrum⑧, the designer in the smaller fashion company will often create designs either by sketching or draping, rather than selecting material from stocks⑨ of fabrics. At this level, the designer may choose to work with a team of textile designers. Experimentation is central to the design process and fabric can become the stimulus for the fashion designer, suggesting new shapes and design ideas. Fabric choice and fashion design, as evidenced by the output[22] of small producers, suggest that fabric cost factors tend to be less important here than for larger companies.

1.2 Fashion designer's attitudes towards textiles

Similar to the fashion designer within a small company, at the level of haute couture, the fashion designer's relationship with fabric lies at the core of their creative process Christian Dior, describing his thoughts about material, stated "Fabric not only expresses a designer's dream, but also stimulates his own ideas. It can be the beginning of an inspiration. Many a dress of mine is born of the fabric alone." Donna Karan works with mills to develop fabrics especially for her; she says "The future of fashion lies in fabrics. Everything comes from fabrics". Pamela Golbin in Paris said, everything evolves from the fabric, so your relationship with the fabric will change the outcome[23]. If you choose chiffon[24] or wool[25], the result of each will be different.

时尚公司的规模和设计师所在公司的类别，决定了公司对待面料的态度。例如，占市场份额大的公司实行安全策略，即重复订单。上一个季节成功的面料，在这个季节早期的设计阶段就发出订单。由于这种操作方法，在这些大公司工作的设计师对已经选择好的面料的性能很熟悉，避免了设计中的悬垂过程。悬垂是一种常用的方法，就是了解面料披挂人体上的特征。小时尚公司完全相反，他们给客户提供一些原创的和独特的面料。因此，使用的面料更多样化和很少重复订单，但是要比大公司对面料进行更多的悬垂性能评估。在大公司的设计师能够与面料生产商保持良好的关系，很容易沟通对面料特征的看法，或最新潮流特征等信息，这样导致新面料的生产，使公司的产品得到更新。而小公司主要集中于选择面料和廓型，使产品中暗含个性。时尚设计师根据他们服务的对象选择面料，最终选择的面料是根据客户的需要和是否得体来判断的。

对于规模较大的公司，由于薄利润率的运作模式，成本因素至关重要。因此，这些公司就要在设计和面料采购方面追求有效成本的解决方案。大公司只有从规模经济中赚取利润，这很关键，如果他们想维持市场份额和稳健的利润。相对于大公司，小设计公司的设计师经常要画设计图或立体裁剪进行创新设计，而不是从库房中选面料。因此，小公司设计师经常与面料设计师的团队一起工作。设计过程的核心是试验，面料可能激发设计师设计新样式和产生新的设计理念。小公司的面料选择和时尚设计由毛利证明，表明面料成本不像大公司那样重要。

1.2　时尚设计师对面料的态度

与小时尚公司的设计师类似，在高级时装层面上，时尚设计师与面料的关系处于设计的核心地位。克里斯汀·迪奥（Christian Dior）曾经描述他对面料的看法。

专业词汇

12. order vi. 订单
13. unique adj. 唯一的
14. repeat n. 重复
15. sustain vt. 维持
16. update vt. 更新
17. cost n. 成本
18. margin n. 利润
19. solution n. 解决
20. share n. 份额
21. profit n. 利润
22. output n. 产量
23. outcome n. 结果
24. chiffon n. 雪纺
25. wool n. 羊毛

通用词汇

⑧ spectrum n. 范围
⑨ stock n. 库存

Apart from being very emotive[26], the fashion designer's relationship with material can also be subject to their personal philosophy and beliefs, which they choose to express through fashion. Alternatively⑩, their design philosophy may simply be avant-garde[27] or culturally alien to a market. This pattern tends to be either generation led, resulting in the emergence of cutting-edge revolutionary fashion designers, or driven by some cultural ethos⑪. These designers have what we might term a radical attitude to cloth in that they challenge preconceived ideas of which materials are "allowed" to be used in fashion. Yohji Yamamoto's philosophy of beauty is quite unconventional compared to the traditional approach to fashion.

It seems that if a designer undertakes such an evolution in their creative process, it will in some ways affect their material choices or the ways in which they choose to develop a design idea. For example, in collaboration with the sports label Adidas, Yamamoto integrated art with engineering in his Autumn/Winter 2001-2002 collection. The fact that a constant stream of young radical designers now explore material innovations raises the question as to whether material innovations themselves actually assist the fashion designer in becoming more adventurous.

1.3 Fabric as a signature[28]

Some fashion designers are as famed for the fabric their collections are made from as they are for the actual fashion collections themselves. In effect, the fabrics act as a transmitter[29] for the fashion label, as is the case with Missoni, Burberry, Pucci, Issey Miyake and St. John Knits. Collaborations between fashion designers and textile companies or textile designers, such as that between Paul Smith and Liberty, have benefits for both fashion and textiles. The use of a limited specific textile design or range of designs can provide an extra edge and element of exclusivity⑫ to a fashion designer's collection. A strongly "identifiable" fabric can "stand out" over and above any other fabric design and choices that could be made.

Through this sort of association, the textile provider is able to reach a greater audience through using an established fashion designer's collection as a vehicle⑬ to distribute their fabrics and so spread their name. Fashion designers don't necessarily have to be associated with "named" textiles; Yeohlee Teng has achieved success through simply being renowned for using particularly exquisite[30] fabrics within her collections.

他说：“面料不仅表达一个设计师的梦想，而且激发设计师理念的产生。它是灵感的开始，我的很多服装灵感产生于面料。”唐纳·凯伦（Donna Karan）与工厂一起开发她需要的面料，她说：“未来的时尚在于面料，各种服装来自面料。”巴黎的帕梅拉·哥布林（Pamela Golbin）说：“每件服装由面料演变而来，因此你与面料的关系将改变服装的样式。你选择雪纺或羊毛，其结果不同。”

时尚设计师与面料的关系除了非常情绪化，还可根据其个人哲学和信仰选择面料，并由时尚来表达。另外，他们的设计可能是前卫的或者从文化上与市场格格不入。这种情形多半是新生代，导致新锐革命化时尚设计师的出现或者受到文化社会思潮的驱动。我们称这些设计师对服装采取了一种激进态度，他们借助时尚，通过使用许可的材料来挑战既定的思想。山本耀司的时尚美的哲学与传统方法相比，相当标新立异。似乎是，如果一个设计师在他们的创作过程中逐渐变化，将在某种程度上影响他们对面料的选择，或影响他们选择设计理念的发展方式。例如，山本耀司与运动品牌阿迪达斯合作中，将艺术与技术结合在一起，设计了2001—2002秋冬系列。事实上，现在不断涌现出激进的年轻设计师，探索材料的创新，从而提出了一个问题，是否材料创新本身有助于时尚设计师越来越喜欢冒险。

1.3 鲜明特征的面料

一些设计师的发布会以面料取胜，他们的名声不亚于那些以设计取胜的设计师。事实上，面料是时尚品牌的传播者。例如米索尼（Missioni）、巴宝利（Burberry）、璞琪（Pucci）、三宅一生（Issey Miyake）和圣约翰（St. John Knits）这些品牌都是因面料出名。时尚设计师与面料公司或面料设计师合作，使两者同时受益，例如保罗·史密斯（Paul Smith）和利伯提（Liberty）的合作。利用独特的限量版面料和设计款式，为时尚设计师的产品提供了独特的优势。一种具有很强识别性的面料可以超越其他任何面料的设计和选择。

通过合作，面料供应商利用闻名遐迩的时尚设计师发布会作为传播他们面料的手段，由此他们的名声也得到了传播，从而有更多的人知道他们。时尚设计师没有必要非得和有名的面料公司合作。邓姚莉（Yeohlee Teng）在她的作品中使用特别精美的面料取得了成功，而一举成名。

专业词汇

26. emotive adj. 情绪的
27. avant-garde adj. 前卫的
28. signature n. 签名
29. transmitter n. 传送者
30. exquisite adj. 精致的

通用词汇

⑩ alternative n. 二中择一
⑪ ethos n. 社会思潮
⑫ exclusivity n. 排外性
⑬ vehicle n. 手段

All of this begs the question as to how we perceive fabric as a component of a garment. If we view garment and fabric as separate identities we might also sometimes ascertain⑭ that the material is as important, if not sometimes more important, than the garment itself. Missoni, Burberry and Pucci are evidence that it is possible for a fashion company to be as famed for their fabric as they are for their runway collections. They all use very distinctive and globally recognized signature fabrics. Burberry is renowned for its check[31], Missoni for its multicoloured[32] knitted stripes[33] and Pucci for its printed psychedelic⑮ swirls⑯. It could also be said that these companies use fashion as a carrier to deliver the fabric to the consumer; after all, they receive acknowledgements and accolades time and again for their fabric as much as for their fashion.

When a fabric speaks louder than the garment it is transformed into, which can be the case with the companies outlined above, the scenario begs the question as to which leads which, does fashion lead textiles or do textiles lead fashion?

1.4 Choosing fabrics

When a fashion designer considers how today's lifestyle makes demands on the modern fashion consumer, fabric choice plays an integral part in the design process. Among those designers who have considered lifestyle, and catered for its demands by incorporating features within their fabrics and clothing, are St. John and Issey Miyake. Although on first impression these designers appear to be worlds apart and, granted they certainly are very different from one another in terms of design philosophy, similarities arise between them in their consideration of the needs of the traveler. Miyake and St. John Knits share a similar viewpoint[34] in that clothes should travel well and take the minimum of care[35]. Both of these companies developed their personal design principles in parallel but in different parts of the world, and they remarkably cultivated a similar approach to fashion in that material research was a priority and became an identifiable trait[36] of each of them.

Following the initial inspiration or concept behind a collection, fiber and fabric selections are the next item considered on the agenda[37] and can take up quite a large proportion of the fashion designer's time. As mentioned earlier, this could even include fashion designers consulting with yarn[38] manufacturers and mills, attending yarn and fabric trade fairs. Some fashion designers, when attending these trade fairs, are already quite clear about the look they are aiming to achieve and so will source specifically the sort of fabrics required. Equally, there are designers who get ideas from fairs.

After fabrics have been selected at fairs or through mills for specific designs, the designer goes back to their studio and begins to sample (prototype a garment) rather than toile (develop on a mannequin), in a fabric similar to the real one purchased. This helps the designer to understand how a fabric will behave when used in a particular design. At this stage, other issues can dominate the process, sometimes overriding design decisions. For instance, concerns about fabric price, supply, delivery or minimum order quantities may end up compromising a design. There are also lead times to deal with, which again vary according to the mill that supplies the fabric. Delivery is relative to the type of fabric requested; therefore decisions about prints and fibers need to be made prior to material selection as they take longer to process.

所有这一切，不禁想问，我们如何看待面料是服装的组成部分。如果我们将服装和面料分开看待，有时我们可能承认，面料即使不比服装更重要，起码也和服装一样重要。米索尼、巴宝利和璞琪证明以面料取胜的时尚公司和以 T 台发布取胜的时尚公司一样有名。这三家公司都采用相当独特、具有全球化识别标志的面料。巴宝利以它的格子闻名，米索尼因五彩缤纷的针织条纹闻名，璞琪的迷幻般涡漩印花图案为人们熟知。也可以这么说，这些公司将时尚作为载体，将面料传递给消费者。总之，他们的时尚和面料屡屡得到人们的承认和赞誉。

服装由面料转变而来，如果面料比服装获得更多的声誉，就像上面提到的那些公司，回避不了谁主导谁的问题，是时尚主导面料还是面料主导时尚？

1.4 面料的选择

当时尚设计师考虑现代生活方式如何影响消费者的时尚需求时，在设计过程中面料的选择起着重要作用，这样的设计师有圣·约翰（St. John）和三宅一生（Issey Miyake）。为了满足生活方式需求，他们将面料和服装的特征结合起来。虽然第一印象似乎这些设计师有天壤之别，可以肯定他们的设计理念彼此存在很大差异，但是在考虑旅行者的需求时，他们有相似之处。三宅一生和圣·约翰都认为旅行服装应该具有良好状态，不要太多的关心。两家公司在世界不同地区，但他们各自开发的设计原则相同，最明显的是他们采取了类似的途径，将研究面料放在首位，这就是他们共同的特征。

一场发布会的背后，首先是灵感和理念，接下来的议事日程就是考虑选择纱线和面料，将花去设计师大量的时间。就像前面提到的，设计师可能要去咨询纱线制造商和工厂，参加纱线和面料展销会。一些时尚设计师，在参加展销会时，已经有清晰的概念，因此直奔目标，专门寻找他们需要的面料。同样，有些设计师从展销会获得灵感。

在展销会选好面料或到工厂设计了专门面料以后，设计师回到他们的工作室，不用白坯布（在人台上研制），而是用类似于真实面料，开始制作样品（服装的原型）。这将有助于设计师了解在特定设计中面料的习性。此时，另外的问题左右着进程，有时推翻了设计决定，例如考虑到面料的价格、供应、交货或最少订单量等问题，一种设计可能就此罢休。还要处理提前期问题，不同面料供应商提前期不同。交货时间与所需面料的种类相关，印花和纱线需要在选择材料之前做出决定，因为它们需要较长的生产过程。

专业词汇

31. check n. 方格
32. multicolored adj. 多色彩的
33. stripe n. 条纹
34. viewpoint n. 观点
35. care vt. 照料
36. trait n. 特点
37. agenda n. 议程
38. yarn n. 纱线

通用词汇

⑭ ascertain vt. 确定
⑮ psychedelic adj. 引起幻觉的
⑯ swirl n. 漩涡

Exercises

(1) Understanding the text.

Read the text and answer the following questions.

1) Do material innovations actually assist the fashion designer in becoming more adventurous?

2) Does fashion lead textiles?

3) Do textiles lead fashion?

(2) Building your language.

The following words and expressions can be used to talk about textiles and fashion. Choose the right ones to fill in the blanks in the following sentences. Change the form where necessary.

reveal	redefine	tool	seek
a variety of	range from	immersive	interface with

Textiles connect ____________ practices and traditions, ____________ the refined couture garments of Parisian fashion to the high-tech filaments strong enough to hoist a satellite into space. High-performance fabrics are being reconceived as ____________ webs, structural networks and information exchanges, and their ability to ____________ technology is changing how the human body is experienced and how the urban environment is built. Today, textiles ____________ their capacity to transform our world more than any other material. Textile Futures highlights recent works from key practitioners and examines the changing role of textiles. Recent developments present new technical possibilities that are beginning to ____________ textiles as a uniquely multidisciplinary field of innovation and research. This unit is an important ____________ for any textile practitioner, fashion designer, architect, interior designer or student designer interested in following new developments in the field of textiles, ____________ new sustainable sources, or just eager to discover new works that reveal the potency of textiles as an ultramaterial.

(3) Sharing your ideas.

After learning about textiles and fashion, now it's the time for you to share your knowledge. Please give a three-minute oral report introducing the relationship between textiles and fashion. Try to make full use of what you've learned from this section; selecting the relevant information from the text and using words and expressions in Building your Language exercises.

Unit 2

Clothing

第2单元

服装

Clothing1 is fiber2 and textile3 material
worn on the body. The wearing4 of clothing
is mostly restricted to human beings and is
a feature of nearly all human societies. The
amount and type of clothing worn depends on
physical, social and geographic considerations.
Some clothing types can be gender-specific,
although this does not apply to cross dressers.
1. Clothing, body and fashion
No reason exists for feeling that ones body is
forbidden topic. What might make it seem so
are the associ[illegible] of fe[illegible] has come to carry.
After all, it [illegible] in [illegible] recent times that
the body, a[illegible] ① concealed
for hundre[illegible] revealed.
This proce[illegible] most to be
escalating, [illegible] manifestation of
the body has [illegible] a [illegible] of surprise and
the phenome[illegible] ded with deep
suspicion②.
It must never[illegible] acknowledged that
we relate many of our concepts of nature to the
human body. We refer, for example, to the

Clothing[1] is fiber[2] and textile[3] material worn on the body. The wearing[4] of clothing is mostly restricted to human beings and is a feature of nearly all human societies. The amount and type of clothing worn depends on physical, social and geographic considerations. Some clothing types can be gender-specific, although this does not apply to cross dressers[5].

2.1 Clothing, body and fashion[6]

No reason exists for feeling that ones body is forbidden topic. What might make it seem so are the associations of fear it has come to carry. After all, it is only in very recent times that the body, after being assiduously① concealed for hundreds of years, has been revealed. This process, once begun, seems almost to be escalating. The increasing manifestation of the body has caused a measure of surprise, and the phenomenon is still regarded with deep suspicion②.

It must, nevertheless, be acknowledged that we relate many of our concepts of nature to the human body. We refer, for example, to "the foot" of a mountain and "the neck" of a river isthmus③, and we say that a tree "stands". Michelangelo extended this parallel when he stated, "The man who cannot master the human body, and particularly its anatomy, will never understand the meaning of architecture."

All our ideas of proportion[7] are related to the body. Our movements arise out of our sight and our instinctive muscular reactions. Whether we are sleeping or walking, our feeling for rhythm[8] is closely connected with our regular heartbeats and the rise and fall of our breathing.

Clothes assume significance only when they are on the body. When they are hung up in a wardrobe[9] they look pathetically④ helpless; they seem to be voicelessly denouncing the cruelty of the tailor who forced them into their state of sad dependence. To really understand clothes it is necessary first to see the reasoning behind them and then to see them, as it was, in action. Clothes are more than just products of a textile factory or exhibits in a museum: they are artifacts⑤, used by people in all activities of daily life-standing, sitting, dancing, working or dying. Their true significance only becomes apparent[10] when we consider how they are related and adapted to the body. So many different human types exist: thin people, fat people, people with large heads, pin-headed people. But human nature being perverse⑥, styles do not always echo the body framework. If the body does not suit[11] a certain style[12] of dress then it is the clothes and not the body which should be modified.

But the determining factor is really neither the body nor the clothes. It is Fashion. Fashion extends far beyond mere clothes; it is affected by how we stand, sit, smile and how we love and how we hate. It is nothing less than an attempt to unify all the expressive capacities of language, gesture and physiognomy⑦ in a given society.

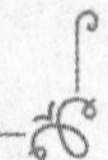

服装就是穿在人身体上的纤维和纺织材料。服装大多数只限制于人类穿着这几乎是人类社会的共同特征。生理、社会和地理因素决定了服装穿着的数量和品种。一些服装样式有性别区分，即不适合男性和女性共同穿着。

2.1 服装、身体和时尚

现在已经没有理由禁止谈论人的身体了。原来之所以禁止，是因为身体似乎总是与一些可能带来害怕不安的事情有联系。身体毕竟已经被严严实实地包裹了几百年，直到近代才被揭开。这个过程一旦开始，就快速地发展。人们对身体越来越暴露有一定程度的惊讶，并且依旧对这一现象持谨慎的怀疑态度。

然而，必须承认，我们将很多自然方面的概念与人的身体对应起来。例如，我们说一座山的“脚”，一条河峡的“颈”，一棵“站立”的树。米开朗基罗（Michelangelo）对此有更深刻的阐述，“人如果不精通人的身体，尤其是身体的解剖，就永远不能理解建筑的含义”。

我们所有比例概念都与身体相关。运动使我们的视野拓宽了，也使我们肌肉反应更加灵敏。不管躺着还是走路，我们感觉到的节奏与我们有规律心跳和呼吸的节奏密切相关。

服装只有穿在身体上时才有意义。当它们被悬挂在衣橱里时显得极其无助。它们似乎在无声地怒吼，残忍的裁缝师逼迫它们处于悲伤孤独的境地。要想真正了解服装，必须先了解它们背后的原因，然后再了解它们，只有这样才是可行的。服装不仅仅是纺织工厂的产品，或博物馆展示的物品。它们是人工制品，人们在日常生活的各种活动中都用到它们——站立、坐着、跳舞、工作或临终。只有当我们考虑它们与身体的关系如何，怎样适应身体的时候，它们的真正意义才变得明显。人类存在各种各样人的体型：瘦的、胖的，头大的、头小的。服装样式不能总是与身体框架一致，往往违背自然人体。如果身体不适合某种着装样式，那么修改的应该是服装，而不是身体。

但是真正的决定因素既不是身体也不是服装，而是时尚。时尚远远超出了单纯的服装，它影响我们如何站、坐、笑和哭，以及爱和恨。它设法在特定社会里统一所有语言、姿态和面貌的表达能力。服装表现为某个时期和环境中产生的艺术形式，所以它不比其他艺术的创造性差。服装设计工艺比得上建筑，但没有建筑的永久性。

专业词汇

1. clothing n. 服装
2. fiber n. 纤维
3. textile n. 纺织品
4. wear vi. 穿着
5. dresser n. 衣着者
6. fashion n. 时尚
7. proportion n. 比例
8. rhythm n. 节奏
9. wardrobe n. 衣橱
10. apparent adj. 易看见的
11. suit v. 适合
12. style n. 风格

通用词汇

① assiduously adv. 勤勉地
② suspicion n. 怀疑
③ isthmus n. 地峡
④ pathetically adv. 哀婉动人的
⑤ artifact n. 人工制品
⑥ perverse adj. 堕落的
⑦ physiognomy n. 人相学

Clothes represent an art form rising out of a period and environment and as such are no less valid than other artistic creations. Dress design[13] is a craft comparable with architecture, but lacking the latter permanence. It is an art which, like music, is in constant movement but, unlike music, is cannot express direct emotion. As are both there arts, dress design is also non-figurative. It seeks not to pretend, but rather to display. It is at one and the same time fettered⑧ and free, as is all genuine art. There is the simple type of dress which innocently declares its purpose, like a ball-dress[14], and then there are carefully constructed complex compositions like the armour[15] of the Middle Ages.

As do all other works of art, clothes reflect the times in which they were created. People reveal themselves unconsciously both in the art forms they accept, and those they reject. No form of art is more subtle than the variations that dress design creates upon that most fascinating of all themes-the human body.

When we consider the subject of dress through the centuries, it is not details of the dressmaker's[16] skill which concern us mod, but rather a gradual awareness of fashion as a camouflage, as a long series of device for hiding the true nature of men and women, created by God in his own image. The value of such duplicity⑨ is surely negative, on a par with the conjuror who draws attention to his empty hand which does not do the trick.

The history of dress is full of deception⑩ and self-delusion, deliberate mistakes and unconscious mistakes, the calculated and uncalculated in a long and illogical record of human folly. In short, it is simply the history of mankind. The interplay between the two sexes, with the whole range of shifting emotion that it encompasses, forms a mirror in which is reflected the ever-changing world of fashion. Dress does not merely show how men and women with to appear; it provides answers to many questions and also a criticism of the people who ask them. Women's fashion through the ages can provide a mass of silent evidence against the tyranny of the male at various times, and men's fashion can make out an equally strong case against the female sex. These situations occur because each sex reacts in accordance with the demands of the opposite sex. In a subtle way, the nature of one sex is revealed in the concealments of the other.

The history of dress cannot be treated as a conducted tour in which wheels revolve and carry both author and reader to a predetermined⑪ destination. The reader must be a pedestrian⑫ and use his legs; he must halt continually to adjust his bearings and change his point of view, sometimes to regard woman from the man's angle and at other times vice versa. It is necessary to observe and study all aspects of the history of dress from both these standpoints before we can hope to gain a real understanding of fashion.

它像音乐艺术，不断运动，但不能直接表达情感。它们两者都是艺术，但服装设计非具象，它不寻求伪装而是展示。它将束缚和自由集于一体，是真实的艺术。像舞会裙等这种简单类型的礼服，很坦白地宣布它的目的，还有一些服装精心构造，例如中世纪的甲胄。

就像所有其他艺术作品那样，服装反映了它被创作的年代。人们无意识地揭示了他们接受和反对的艺术形式。没有一种艺术形式像服装设计那样，用各种微妙的变化，为人体创建最迷人的主题。

当我们思考几个世纪以来的具体服装时，不是裁缝师的技术细节与我们的时尚相关，而是对时尚渐渐产生意识，可以作为一种伪装和一系列装置来掩盖男性和女性的真实自然特征，即上帝根据自身形象创造的人。这种表里不一无疑起着负面作用，就像魔术师将注意力吸引到他没有做诡计的空手上。

着装历史充满了欺骗和自我欺骗、故意犯错和无意识失误、精心策划或一时冲动，记录了长期以来人类愚蠢的和不符合逻辑的行为，简单地说，就是人类一部简单的历史。有史以来夹带着情感变化的两性之间的相互作用，构成一面镜子，透过它映射出永远变化的时尚世界。服装不仅显示男女的外貌如何，还给予了很多问题的答案和对提问题的人的批判。在历史的不同时期，女性时尚提供了大量的反对男性独裁的无声证据；而男性时尚也显示了强烈反对女性性别的对等情形。这些情况的发生反映了一种性别对另一种性别的要求。以微妙方式，一种性别特征被揭示时，另一种性别特征就被掩盖。

对待着装历史不能像对待导游团那样，转动车轮将作者和读者带到一个事先决定的目的地。读者必须是一个使用双腿的徒步者，不停地止步，调整态度和改变观点，有时从男性的角度看待女性，有时则相反。必须从男女两性观察和研究着装的历史，我们才有希望获得对时尚的真正了解。

专业词汇

13. design vt. 设计
14. ball-dress n. 舞会服
15. armour n. 甲胄
16. dressmaker n. （女装）裁缝

通用词汇

⑧ fetter vt. 束缚
⑨ duplicity n. 欺骗
⑩ deception n. 瞒骗
⑪ predetermine vi. 注定
⑫ pedestrian n. 行人

2.2 The uses of clothing

2.2.1 Utility

Clothing has evolved to meet many practical and protective purposes. The environment is hazardous, and the body needs to be kept at a mean temperature to ensure blood circulation and comfort. The bushman needs to keep cool, the fisherman to stay dry; the firemen needs protection from flames and the miner from harmful gases.

Humans have shown extreme inventiveness in devising clothing solutions to environmental hazards. Some examples include: space suits, air conditioned clothing, armor, diving suits, swimsuits[17] bee-keeper gear, motorcycle leathers[18], high-visibility clothing, and other pieces of protective clothing. Meanwhile, the distinction between clothing and protective equipment is not always clear-cut, since clothes designed to be fashionable often have protective value and clothes designed for function often consider fashion in their design.

2.2.2 Modesty

We need clothing to cover our nakedness[19]. Society demands propriety and has often passed sumptuary (clothing) laws to curb⑬ extravagance and uphold decorum⑭. Most people feel some insecurity about revealing their physical imperfections, especially as they grow older; clothing disguises and conceals our defects, whether real or imagined. Modesty⑮ is socially defined and varies among individuals, groups and societies, as well as over time.

In many Middle Eastern countries a debates still rages between liberals and fundamentalists as to how covered a women should be, and in many contemporary societies women still wear long skirts[20] as a matter of course. Europeans are generally less inhibited than Americans, but the trend for "casual Fridays" and dressing down office has been imported from the United States.

2.2.3 Sexual attraction[21]

Clothing can be used to accentuate⑯ the sexual attractiveness and availability of the wearer. The traditional role of women as passive sexual objects has contributed to the greater eroticization[22] of female clothing. Eveningwear[23] and lingerie[24] are made from fabrics that set off or simulate the texture[25] of skin. Accessories[26] and cosmetics[27] also enhance allure. Many fashion commentators and theorists have used a psychoanalytic⑰ approach, based on the writing of Sigmund Freud and Carl Jung, to explain the unconscious processes underlying changes in fashion.

2.2 服装的使用

2.2.1 效用性

服装已经发展到能够满足很多实际和保护的用途。在危险的环境，身体需要保持均衡的温度以保证血液循环和舒适。人身体需要保持凉爽，渔夫需要保持干燥，消防员需要防火，矿工需要预防有毒气体。

人类在设计服装解决危险的环境问题方面，已经显示出极大的创造性。一些例子包括宇航服、空调服、防弹服、潜水服、游泳服、防蜂服、摩托车皮服、高度可视服装和其他保护性服装。然而，服装和保护性设备的区别不是很清晰，因为时髦设计的服装往往具有保护性能，设计功能性服装经常考虑时尚性。

2.2.2 遮羞性

我们需要服装覆盖我们的裸体。社会需要得体，经常颁布奢侈禁令限制奢侈和维持体统。大多数人感到揭示自身身体的瑕疵有点不自信，特别是他们逐渐变老以后。服装伪装和掩盖我们真实或假想的缺陷。端庄是社会性定义，因不同的个体、群体和社会以及不同时代而变化。

在中东很多国家，在自由主义者和原教旨主义者之间仍然激烈地争论女性应该如何着装。在当代许多社会中，女性穿长裙，仍然被认为理所当然。欧洲人的着装限制渐渐地比美国人少，但“休闲星期五”和脱下办公服来源于美国。

2.2.3 性吸引

服装可用来强调穿着者的性吸引力和性能力。女性是被动的性对象的传统角色使女性服装更加性感。晚装和内衣采用的面料引起皮肤触感或模拟皮肤的肌理。配饰和化妆品也增强诱惑力。许多时尚评论家和理论家根据西格蒙德·弗洛伊德（Sigmund Freud）和卡尔·荣格（Carl Jung）的理论，采用了心理分析方法，解释时尚变化的潜意识过程。

专业词汇

17. swimsuit n. 泳装
18. leather n. 皮革
19. nakedness n. 裸体
20. skirt n. 裙子
21. attraction n. 吸引
22. eroticization n. 性感
23. eveningwear n. 晚装
24. lingerie n. 贴身内衣
25. texture n. 纹理
26. accessory n. 配饰
27. cosmetic n. 化妆品

通用词汇

⑬ curb vt. 限制
⑭ decorum n. 端庄得体
⑮ modesty n. 端庄
⑯ accentuate vt. 强调
⑰ psychoanalytic adj. 心理分析的

The concept of the "shifting erogenous⑱ zone" (developed by John Carl Flügel (1874-1955), a disciple⑲ of Freud, in about 1930) proposes that fashion continuously stimulates sexual interest by cycling and focusing the attention on different parts of the body for seductive⑳ purposes, and that a great many articles of clothing are sexually symbolic of the male or female genitals. From time to time overtly sexualized clothing, such as codpiece[28] or the brassiere[29] comes into vogue[30].

2.2.4 Adornment[31]

Adornment allows us enrich our physical attractions, assert our creativity and individuality, of signal membership or rank within a group or culture. Adornment can go against the needs for comfort, movement and health, as in foot-binding, the wearing of corsets[32] or piercing[33] and tattooing[34]. Adornments can be permanent or temporary, additions to or reductions of the human body. Cosmetics and body paint, jewelry[35], hairstyling and shaving[36], false nails[37], wigs[38] and hair extensions, suntans, high heels[39] and plastic surgery are all body adornments. People generally, and young women in particular, attempt to conform to the prevailing ideal of beauty. Bodily contortions㉑ and reshaping through foundation garments[40], padding[41] and binding have altered the fashionable[42] silhouettes throughout the ages.

2.2.5 Symbolic differentiation

People clothing to differentiate and recognize profession, religious affiliation㉒, social standing or lifestyle. Occupational dress is an expression of authority and helps the wearer stand out in a crowd. The modest attire[43] of a nun announces her beliefs. In some countries, lawyers and barristers cover their everyday clothes with the garb[44] of silk[45] and periwig in order to convey the solemnity㉓ of the law. The wearing of designer labels or insignias㉔, and expensive materials and jewelry, may start as items of social distinction, but often trickle down through the strata㉕ until they lose their potency as symbols of differentiation.

2.2.6 Social affiliation

People dress alike in order to belong to a group. Those who do not conform to the accepted styles are assumed to have divergent ideas and are ultimately mistrusted and excluded. Conversely, the fashion victim, who conforms without sensitivity to the rules of current style, in perceived as being desperate to belong and lacking in personality and taste. In some cases clothing is a statement of rebellion against society of fashion itself. Although punks[46] do not have a uniform[47], they can be recognized by a range of identifiers: torn clothes, bondage[48] items, safety pins[49], dramatic hairstyles[50] and so on.

“性感带转移”的概念［由弗留葛尔（John Carl Flügel）（1874—1955）大约于1930年提出，他是弗洛伊德的追随者］是指，时尚不断地循环和聚焦身体的某个部位，激起性感兴趣，以达到诱惑目的，服装的很多部件是男性或女性生殖器的象征，不时地有明显性感化服装的出现，例如男性的裆布或女性文胸成为时尚。

2.2.4 装饰

装饰可以丰富我们身体的吸引力，体现我们的创造力和个性，象征某个集团或文化中的成员和地位。装饰可以违背舒适、运动和健康的需求，例如裹脚、紧身胸衣，穿刺和文身。装饰可以是永久性或临时性的，对身体进行添加或减少。化妆和体绘、首饰、发型和刮胡须，假指甲、假发和植发、晒黑、高跟鞋和整形手术等都是身体装饰。一般来说，人们尤其是年轻女性设法顺应流行的理想美。不同年代有不同的时尚廓型，它们的产生是通过基础服装、衬垫和捆绑使身体扭曲和重新塑造。

2.2.5 象征性差异

人们通过着装认识和区分职业、宗教信仰、社会地位或生活方式。职业装是职权的表达，有利于穿着者在人群中脱颖而出。尼姑的得体服装宣告她的信仰。在某些国家，律师和法律顾问每天穿丝绸长袍和戴假发，为了表达法律的尊严。穿戴名牌或徽章、昂贵材料和饰品，可以作为社会区分的物品，但是经常是由上向下流向社会底层，直到它们失去差别象征的效用。

2.2.6 社会附属关系

人们穿着相同服装为了归属群体。那些与已有风格着装不一致的人，被假定为具有不同的想法，最终不被信任，并被群体排除。相反，时尚牺牲者只会顺从，没有对现时时尚规律的敏感性，被认为极度渴望归属，缺乏个性和品位。在某些情形下，反对时尚的着装是反抗社会的声明。朋克尽管没有制服，他们有很多特征被认知：撕坏的服装、绑带、安全别针、奇特的发型，等等。

专业词汇

28. codpiece n. 遮阴布
29. brassiere n. 文胸
30. vogue n. 时尚
31. adornment n. 装饰
32. corset n. 紧身胸衣
33. pierce vt. 刺穿
34. tattoo n. 文身
35. jewelry n. 首饰
36. shave vt. 剃须
37. nail n. 指甲
38. wig n. 假发
39. heel n. 后跟
40. garment n. 服装
41. pad vt. 衬垫
42. silhouette n. 轮廓
43. attire n. 服装
44. garb n. 装扮
45. silk n. 丝绸
46. punk n. 朋克
47. uniform n. 制服
48. bondage n. 绑带
49. pin n. 别针
50. hairstyle n. 发型

通用词汇

⑱ erogenous adj. 唤起情欲的
⑲ disciple n. 信徒
⑳ seductive adj. 诱惑的
㉑ contortion n. 扭曲
㉒ affiliation n. 附属
㉓ solemnity n. 庄严
㉔ insignia n. 证章
㉕ strata n. 社会阶层

2.2.7 Psychological self-enhancement

Although there is social pressure to be affiliated to a group, and many identical garments and fashions are manufactured and sold through vast chain stores, we rarely encounter two people dressed identically from head to toe[51]. While many young people shop with friends for help and advice, they do not buy the same outfits. Whatever the situation, individuals will strive to assert their own personal identity through the use of make-up[53] hairstyling and accessories.

2.3 Different words

Attire "means": (1) clothing of a distinctive style or for a particular occasion; (2) put on special clothes to appear particularly appealing and attractive.

Apparel[54] is very often a ceremonial type of clothing.

"Garment" refers to items of clothing in an almost official way-often linked to the word list when describing clothing belonging to someone.

"Grab" is akin to apparel and is very often associated with a form of clothing that's usual or worn to disguise the person inside them.

"Clothing" is the simplest form of all to describe a collection of clothes. "Garment" and "Apparel" are not identical. They may sometimes be used synonymously when used as adjectives, but when used as nouns they differ grammatically, for example:

——The word "Garment" refers to a single piece of clothing. It is a countable noun.

——The word "Apparel" refers collectively to clothing (and thus usually refers to more than one piece). It is an uncountable noun.

There are also differences in which words typically collocate with them when they modify[55] another noun. For example:

——a garment bag

——an apparel store

The term "Costume[56]" can refer to wardrobe and dress in general, or to the distinctive style of dress of a particular people, class, or period. "Costume" may also refer to the artistic arrangement of accessories in a picture, statue, poem, or play, appropriate to the time, place, or other circumstances represented or described, or to a particular style of clothing worn to portray the wearer as a character or type of character other than their regular persona at a social event such as a masquerade[57], a fancy dress party or in an artistic theatrical performance.

One of the more prominent places people see costumes is in theatre, film and on television. In combination with other aspects, theatrical costumes can help actors portray characters' age, gender role, profession, social class, personality, ethnicity, and even information about the historical period/era, geographic location and time of day, as well as the season or weather of the theatrical performance. Often, stylized theatrical costumes can exaggerate some aspects of a character.

National costume or regional costume expresses local (or exiled) identity and emphasizes a culture's unique attributes. It is often a source of national pride. Examples of such are a Scotsman in a kilt[58] or a Japanese person in a kimono[59].

2.2.7 心理自我提升

尽管人们有隶属于群体的社会压力，许多相同的服装和时尚被制造出来，在各个连锁店销售，但我们很少碰到两个人从头到脚穿得完全一样。许多年轻人和朋友一起逛商店，可以获得帮助和建议，他们不会购买相同的整套服装。不管什么情形，个人总是通过化妆、发型和配饰来维护他们的个性特征。

2.3 不同的词

“Att”意思：（1）独具风格或者特定场合的服装；（2）穿上特别的服装显得特有吸引力和魅力。

“Apparel”经常是指一种礼仪性服装。

“Garment”是一种正式的用法，经常在列举服装属于某人时用这个词。

“Garb”与“Appare”相似，通常指独特的服装，或用来掩盖人内心世界的服装。

“Clothing”是描述服装总称的词汇中形式最简单的词。“Garment”和“Apparel”的不同之处在于，当用作形容词时它们是同义词，但是作为名词时，它们在语法上不同，例如：

“Garment”指单独一件衣服，是可数名词。

“Apparel”指服装的整体（因此，通常指服装多于一件），是不可数名词。

当它们与其他词搭配或修饰其他名词时，也不相同。例如：

一个“服装”袋，一家“服装”商店。

“Costume”一般指衣橱和着装，或者指特定的人、阶层或时期的特定风格的服装。“Costume”也可以指在绘画、雕塑、诗歌或戏剧中，具有艺术化布置的配件，或者与所表现或描绘的时间、地点或场景相吻合的服装。或者某种特定样式的服装，塑造穿着者在特定社会场合中具有某种特征或某类特征，而不是他们真实的个性，例如化装舞会、一种幻想的着装聚会或一种艺术性戏剧表演。

“Costume”更多地用于剧院、电影和电视上的服装。与其他方面相结合，舞台服装可以帮助演员塑造人物年龄、性别角色、职业、社会阶层、个性、种族，甚至有助于体现戏剧表演中讲述的历史时期时代、地理位置和时间、季节和天气。通常情况下，独具风格的舞台服装可以夸张人物的某些特征。

民族服装和地域服装表达了地域（或流亡）特征，强调了一种独特的文化性往往是民族自豪感的源泉，例如苏格兰褶裥短裙和日本和服。

专业词汇

51. toe n. 脚趾
52. outfit n. 全套服装
53. make-up n. 化妆
54. apparel n. 服装
55. modify vt. 修改
56. costume n. 戏装
57. masquerade n. 化装舞会
58. kilt n. 褶裥短裙
59. kimono n. 和服

Exercises

(1) Understanding the text.

Read the text and answer the following questions.

1) Michelangelo said, "The man who cannot master the human body, and particularly its anatomy, will never understand the meaning of architecture." What is your understanding to it?

2) Based on the study of all aspects of history of dress, how can we gain a real understanding of fashion?

3) National costume or regional costume expresses local (or exiled) identity and emphasizes a culture's unique attributes. Can you list more examples?

(2) Building your language.

The following words and expressions can be used to talk about Clothing. Choose the right ones to fill in the blanks in the following sentences. Change the form where necessary.

outlast	sustainability	counterparts	worst
notion	sturdily	measure	durable

If you've been shelling out for designer clothing in the hopes that it will be more ________ than cheaper options, you might want to reconsider. A new study from The University of Leeds suggests that low-cost clothing might actually ________ pricier pieces. The study was led by Mark Sumner, a lecturer in fashion and ________. Using samples of T-shirts and jeans from a variety of brands and price points, researchers ________ factors such as seam strength, colorfastness, and how long it took for the fabric of each piece to develop rips or tears. Despite the popular ________ that designer clothes are more ________ constructed than off-the-rack items, the study found that fast-fashion T-shirts and jeans usually out-performed their luxury ________. "Some of the garments performed very well across a wide range of tests-more often than not, the best products were 'fast-fashion' products." Sumner told The Telegraph. According to Sumner, designer labels T-shirts were "the ________ performing" in all of the tests.

(3) Sharing your ideas.

Please write a short introduction (of around 300 words) to introduce fashion commentators Sigmund Freud and Carl Jung and his essential ideas. Try to make full use of what you've learned from Text, including the relevant information from the reading text as well as words and expressions.

The human eye can discern ② 350,000 colors but the human memory for color is poor. Most people cannot remember a specific color for more than a few seconds. In every day conversations about color, it is sufficient to refer to a few colors by name red, yellow green, blue, white, and black and add a qualifier light1or dark2, bright3 or dull4, cool5 or warm6. General terms are not sufficient for communicating color information for design and manufacturing. Exact identification, matching7, and reproduction of colors require an effective system with colors [illegible]ged [illegible]uential order and identified [illegible]s. Systems such as th[illegible] basic three characteristics of co[illegible]saturation9, and value10 (Figure 6.1[illegible] Hue refers to the [illegible]ach color system designates a set of b[illegible]rs. Varying the other two characterist[illegible]s out the system. Saturation (also call[illegible] intensity 12chroma13) refers to the strength [illegible] purity14 of color and value to the lightness or darkness of the color.

Unit 3 Color

第3单元 色彩

Color grabs[①] customers' attention, makes an emotional connection, and leads them to the product. Even when the basic product stays the same, changing the color gives a sense of something new.

The human eye can discern[②] 350000 colors, but the human memory for color is poor. Most people cannot remember a specific color for more than a few seconds. In every day conversations about color, it is sufficient to refer to a few colors by name-red, yellow, green, blue, white, and black-and add a qualifier—light[1] or dark[2], bright[3], or dull[4], cool[5] or warm[6]. General terms are not sufficient for communicating color information for design and manufacturing. Exact identification, matching[7], and reproduction of colors require an effective system with colors arranged in sequential order and identified with numbers and letters. Systems such as this are based on the basic three characteristics of color-hue[8] saturation[9], and value[10] (Figure 3. 1).

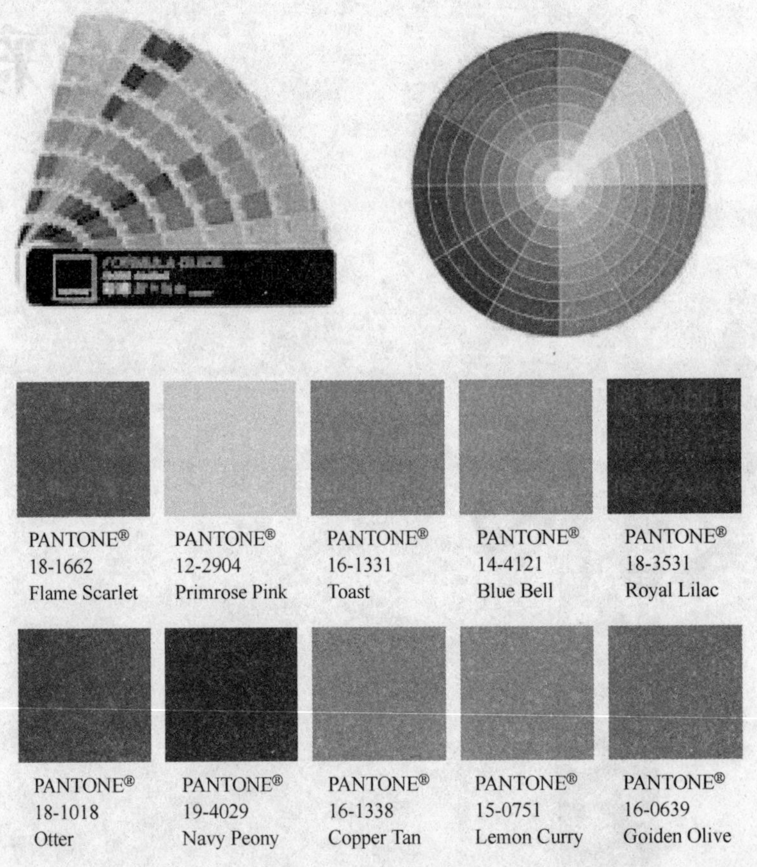

Figure 3. 1 The Pantone Professional Color System

色彩能够吸引客户的注意，使产品与情感联系起来。甚至当产品基本相同时如果改变色彩，会呈现出新的感觉。

人的眼睛能够辨别350000种色彩，但是，人的色彩记忆很差。大多数人记得某种色彩不超过几秒钟。在日常生活中讨论的色彩，几种就足够了，它们的名字是红、黄、绿、蓝、白和黑，再加一些修饰词——浅和深色、明和暗色、冷和暖色。在设计和制造时，一般的术语不够用来交流色彩信息。准确地识别、匹配和生产色彩需要有效的体系，将色彩按照一定的序列进行排列，根据数字和字母进行识别。这种体系基于色彩的三个基本特征：色相、饱和度和明度（见图3.1）。

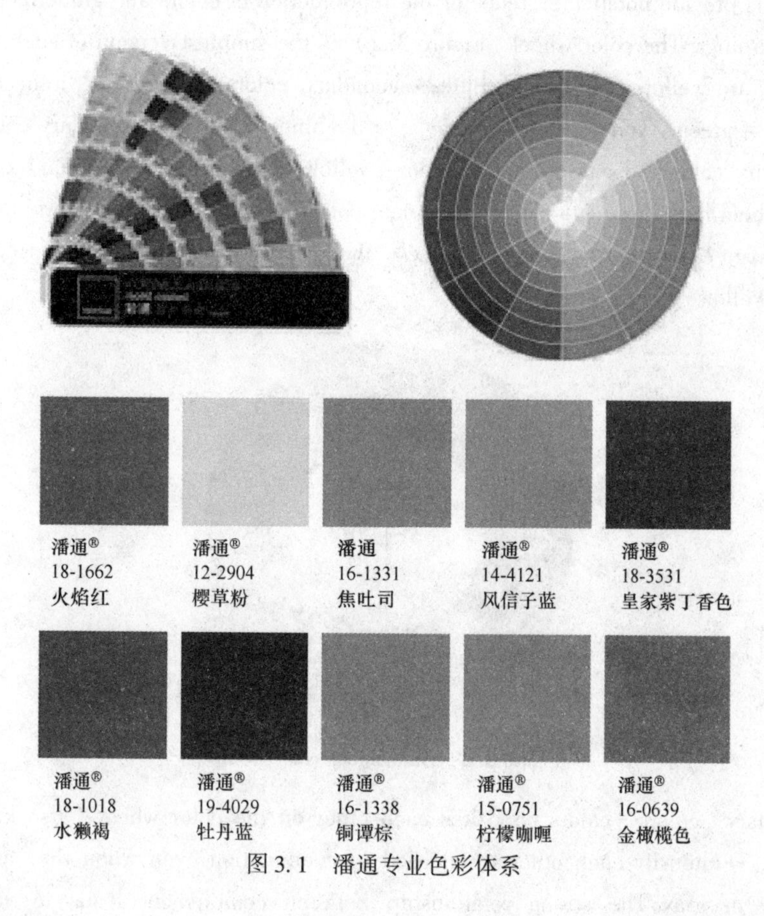

图3.1 潘通专业色彩体系

专业词汇

1. light adj. 淡色
2. dark adj. 深色
3. bright adj. 亮色
4. dull adj. 暗色
5. cool adj. 冷色
6. warm adj. 暖色
7. match vi. 匹配
8. hue n. 色相
9. saturation n. 饱和度
10. value n. 明暗

通用词汇

① grab vt. 夺取

② discern vt. 辨明

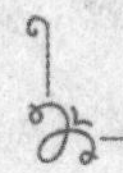

Hue refers to the color-each color system designates a set of basic colors. Varying the other two characteristics[11] fills out the system.

Saturation (also called intensity[12] or chroma[13]) refers to the strength or purity[14] of color and value to the lightness or darkness of the color. The term tint applies to any color when white is added. Shade[15] refers to colors mixed with black. Tone[16] describes a grayed color. Creating a new tint, shade, or tone does not change the designation of the hue, but it does change the value and intensity of a color.

Color systems provide notation systems for the reproduction of colors and guidelines for harmonious[17] color groupings. The color wheel (Figure 3.2) is the simplest version of such a system. The primary colors are yellow, red, and blue; secondary colors are mixed[18] from two primaries (yellow +blue= green, yellow +red=orange[19], red +blue = violet[20]). Tertiary colors are mixed from one primary color and one secondary color (yellow + green= yellow-green). Relationships on the color wheel help designers select coordinate colors. When a patterned fabric or an ensemble uses colors next to each other on the color wheel, the color scheme is called analogous—for example, yellow, yellow-green, and green.

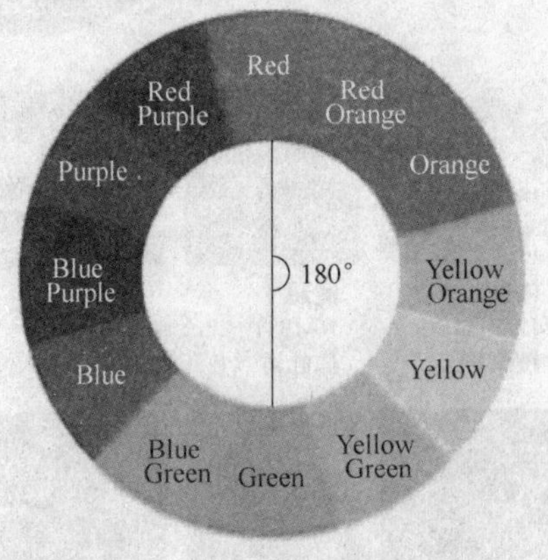

Figure 3.2 Harmonious color groups

Complementary③ colors—colors opposites each other on the color wheel (also known as color complements) —intensify each other when used in combination even when they are mixed with white, black, or gray. The strong relationship between complement leads to variations: the double complement (two sets of complementary colors) and the split complements (a color plus the two colors on either side of its complement).

Three colors spaced equidistant on the color wheel are called a triad④—red, yellow, and blue make the primary triad; Green, orange, and violet make the secondary triad. An infinite number of color combinations can be developed by varying the value and intensity of the colors.

There are two kinds of color systems in American:

- The Munsell Color System, which includes a color atlas, the Munsell Book of Color (1976), with about 1600 chips arranged in equal steps of hue, value, and chroma (intensity or saturation) and a notation for each.

色相就是颜色的名称。每个色彩体系指定了一组基本的色彩。变化色彩的其他两个特征就构成完整性的色彩体系。

饱和度（也称为密度或浓度），是指色彩的力度或纯度。而明度是指色彩的淡或深。浅色调是指添加了白色以后的任意色彩。阴影是指混合黑色后的色彩。调子用来描述加灰的色彩。创造新的浅色调、阴影和调子不改变色相，但是改变了色彩的明暗和浓度。

色彩体系是一种标记体系，用于色彩的再生产，对取得和谐的色彩群组有指导作用。色轮是一种最简单形式的体系，如图 3.2 所示。原色是黄、红和蓝；间色是两种原色的混合（黄+蓝=绿，黄+红=橙，红+蓝=紫）。第三色是一种原色和一种间色的混合（黄+绿=黄绿）。色轮上的色彩关系有助于设计师选择协调的色彩。面料图案或色彩组合时，使用色轮上彼此靠近的色彩，这种色彩方案叫作同类色，例如，黄、黄绿和绿。

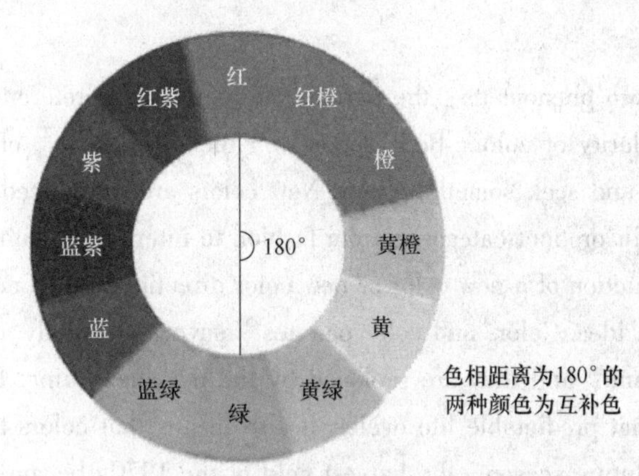

图 3.2　色彩体系提供了色彩和谐群组的指导原则

补色——在色轮上相对的色彩（也称为色彩的补色）——当它们组合在一起时浓度要相等，甚至在混合白、黑或灰色时也要使浓度一致。补色之间的强度关系产生变量：双互补色（两个互补的色彩）和分裂互补（一种色彩与它补色两边的色彩构成的互补）。

在色轮上构成等边的三种色彩称为三色组合——红、黄和蓝构成三原色组合；绿、橙和紫色构成间色的三色组合。通过改变色彩的纯度和浓度可以得到无穷的色彩组合。

在美国有两种普遍的色彩体系：

· 蒙赛尔色彩体系，*Munsell Book of Color*（*1976*）一书中有 1600 块色彩图标集，它们按照色相、纯度和浓度（密度和饱和度）等距离的排列，每一种色彩都有标注。

专业词汇

11. characteristic adj. 特征
12. intensity n. 纯度
13. chroma n. 色度
14. purity n. 纯度
15. shade n. 阴影
16. tone n. 色调
17. harmonious adj. 和谐的
18. mix v. 混合
19. orange n. 橙色
20. violet n. 紫罗兰

通用词汇

③ complementary adj. 互补的
④ triad n. 三个一组

· The Pantone ® Professional Color System, which includes a color atlas, The Pantone Book of Color (1990), with 1225 colors identified by name and color code. A color notation system may look incomprehensible at first glance, but to color professionals the system becomes a precise language for color identification. The complete Munsell notation is written symbolically: H(ue) V(alue)/C(hroma) . For a vivid red, the notation would read 5R 6/14. For finer definitions, decimals are used5. 3R 6. 1/14. 4.

Beginning with 1700 hues, the company expanded the range by 56 colors in 1998 and by 175 colors in 2002 in an attempt to provide a comprehensive mix of colors for fashion, architecture and interior design, and industrial design.

3. 1 Color cycles

Color cycles refer to two phenomena: the periodic shifts in color preferences and the patterns of repetition[22] in the popularity of colors. Both depend on the mechanism[23] of boredom—people get tired of what they have and seek something new. New colors are introduced to the marketplace[24], available to consumers in product categories from fashion to interiors to automobiles. There is a lag time between the introduction of a new color or new color direction and its acceptance while people gain familiarity with the idea. Colors and color palettes[25] move from trendy to mainstream. In time, interest in the colors wane, and they are replaced by the next new thing. This mechanism means that colors have somewhat predictable life cycles. It also means that colors that were once popular can be positioned in a future season—the harvest gold of the 1970s became the sunflower gold of the 1990s.

The vogue for a group of colors evolves over a period of 10 to 12 years, reaching its peak in mid-cycle. The color usually appears in fashion, moving quickly to high-end Interiors, and then to the rest of the consumer products market.

3. 2 Long-wave cycles

The recurrences of color themes can be traced not just in decades but in centuries. In the Victorian era, Owen Jones chose rich, bright, primary colors for London's Crystal Palace in 1851. He defended his choices saying that these same colors had been used in the architecture of the ancient Greeks. To some color historians, this comment illustrates the cycling of color through history, specifically the periodic return to primary hues. The cycle begins with bright, saturated, primary colors; this is followed by an exploration of mixed, less intense colors; then, it pauses in neutral[27] until the rich, strong colors are rediscovered.

Researchers have confirmed a periodic Swing (Figure 3. 3) from high chroma colors, to "multicoloredness", to subdued[28] colors, to earth tones, to achromatic⑤ colors (black, white, and gray), and back to high chroma. In the period between 1860 and the 1990s, there were four marked color cycles lasting between 15 and 25 years.

·潘通专业色彩体系，包含了色彩图集。*The Pantone Book of Color*（*1990*）一书中有1225块色彩图标集，由名称和标号来识别它们。乍一看，色彩的符号体系看上去十分难懂，但是对于色彩专业人士来说，这种体系成了识别色彩的精确语言。一种色彩完整地表示为：H(ue) V(alue)/C(hroma)。例如，鲜艳的红色表示为5R6/14。要想更加精确的定义，可以使用小数点5.3 R6.1/144。

潘通公司于1998年在起初的1700种色彩上增加了56种，2002年又增加了175种。目的是努力为时尚、建筑和室内设计提供广泛的色彩搭配。

3.1 色彩周期

色彩周期是指两种现象：偏爱的色彩周期性变化和色彩流行模式的反复。两者都是因为厌倦心理所致——人们对拥有的东西感到疲倦，寻求新的东西。新的色彩被引入市场，然后到达消费者，体现在时尚、室内设计、汽车等诸多产品上。在新色彩或新色彩趋势引入后，到人们熟悉这种理念并接受它，有一段时间的滞后性。色彩和色彩调色板从潮流进入到主流，同时，当一种色彩的兴趣减退时，就有另一种新的色彩替代它，这种机制说明色彩的生命周期可以预测，且曾经流行的色彩在未来季节里将再次流行。20世纪70年代丰收金色在90年代变成了向日葵金色。

一组色彩演变的时尚周期为10～12年，在周期中间阶段达到高峰。色彩通常首先出现在时尚中，很快传播到高端室内用品，然后到达其他消费产品市场。

3.2 长周期

色彩主题的回归可能追溯到几个年代以前，甚至几个世纪以前。维多利亚时代，欧文·琼斯（Owen Jones）在1851年为伦敦的水晶宫选择了丰富亮丽的原色色彩。他为自己的选择争辩到，这些色彩与古希腊建筑中的色彩相同。对于一些色彩历史学家来说，这种评论注释了色彩的循环具有历史性，特别是周期性地复苏原色。周期开始于明亮、饱和的原色，然后就探索混合色，减少色彩的浓度，接下来中性色，直到丰富、浓烈的色彩再次出现。

研究者已经证实色彩的周期性回转，如图3.3所示。从高浓度色彩、多彩色、柔和色、泥土色，到无彩色（黑、白和灰），再回到高浓度色彩。从1860年到20世纪90年代，标志着色彩四次循环，持续的时间15～25年。

专业词汇

21. symbolical adj. 表示像征的，符号的
22. repetition n. 反复
23. mechanism n. 机制
24. marketplace n. 市场
25. palette n. 调色板
26. mainstream n. 主流
27. neutral adj. 中性色
28. subdued adj. 柔和色

通用词汇

⑤ achromatic adj. 消色差的

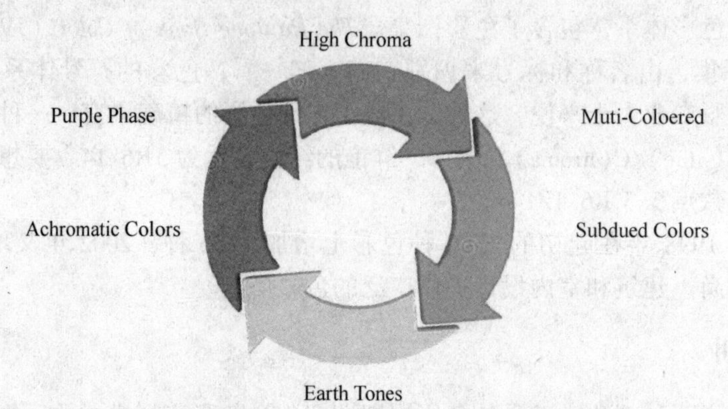

Figure 3. 3　A periodic swing of colors

3. 3　Color cycles and cultural shifts

Color cycles can be sparked by new technology. This happened about ten years after the opening of the Crystal Palace in the mid-1800s. One of the first synthetic[29] dyes[30] was introduced by the French: a color-fast purple[31] called mauve[32]. The color became the rage[33]—Queen Victoria wore the color to the International Exhibition of 1862— and gave its name to the Mauve Decade. Other strong synthetic dyes for red and green soon followed, allowing for a strong color story in women's clothing. Something similar happened in the 1950s when the first affordable[34] cotton[35] reactive dye for turquoise[36] led to a fad for the color and moved it from eveningwear into sportswear.

Economic conditions also disturb color cycles and start new ones. In the depressed 1930s, fabrics and colors were chosen as investments—the buffs[37], grays, and subdued greens and blues were low-key, could be worn more often, didn't show dirt, thus seldom need to be cleaned and, therefore, lasted longer. The steep drop in the stock market in 1987 coincided with the eclipse of bright color as the fashion look of the 1980s and the ascendance of Japanese design featuring austere[38], minimalist[39] black clothing.

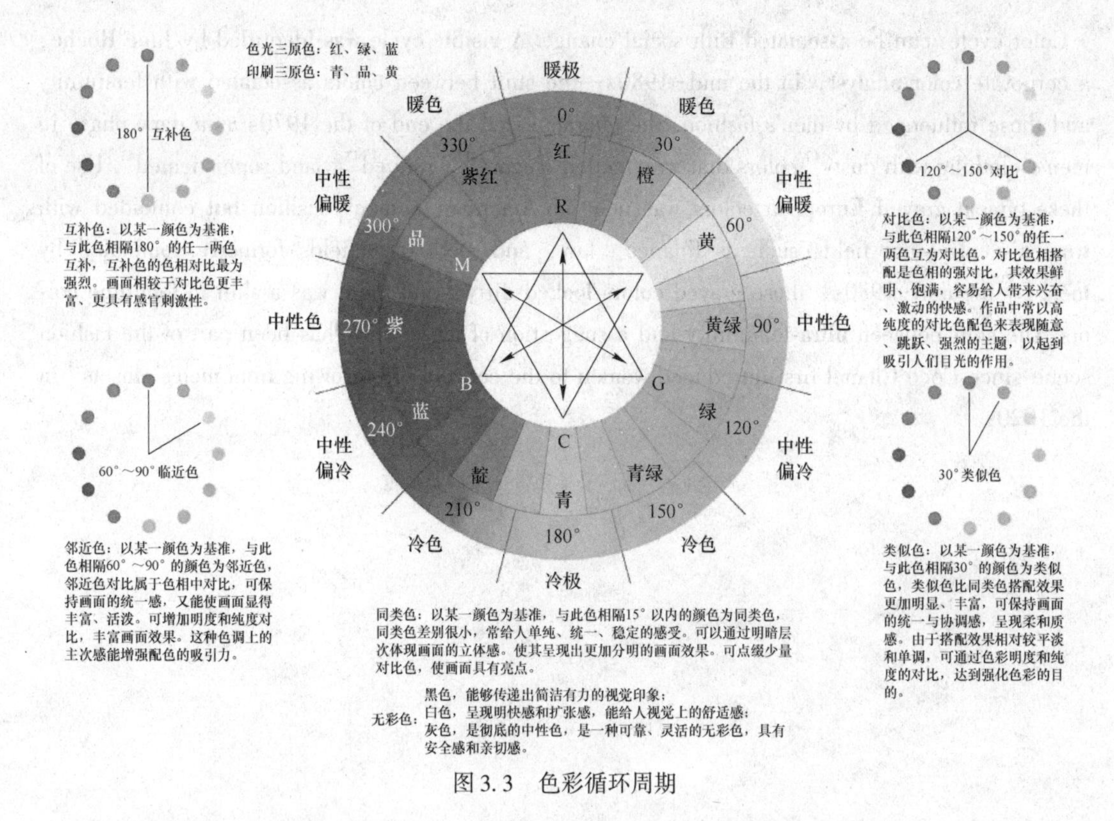

图 3.3　色彩循环周期

3.3　色彩周期和文化变迁

新技术引发色彩变化。这大约发生在 19 世纪中期水晶宫开幕后十年。法国人发明了一种活性紫色彩的合成染料，称之为苯胺紫。这种色彩疯狂流行，被称为苯胺紫年代。1862 年，维多利亚女王穿着这种色彩的服装去参观国际博览会。很快其他浓烈的合成染料，使女性服装大放异彩，如红色、绿色。20 世纪 50 年代也发生了类似的情形，首次出现了价格便宜的棉反应绿松石色彩染料，掀起了一股热潮，从晚装到运动服都采用这种色彩。

经济条件扰乱色彩周期和产生新的色彩。在 20 世纪 30 年代经济萧条时，面料和色彩的选择像投资一样很低调——黄褐色、灰色和暗绿色以及蓝色，经常穿不显脏，因此洗涤次数减少，也能持续很长时间。1987 年急剧下跌的股市应和了 20 世纪 80 年代时尚面貌中亮丽色彩的退却，也应和了日本简朴设计风格和极简主义黑色服装的兴起。

专业词汇

29. synthetic adj. 合成
30. dye n. 染料
31. purple n. 紫色
32. mauve n. 淡紫色
33. rage n. 风靡一时
34. affordable adj. 买得起
35. cotton n. 棉花
36. turquoise n. 蓝绿色
37. buff n. 米色
38. austere adj. 朴素的
39. minimalist adj. 极简主义

Color cycles can be associated with social change. A visible cycle was identified by June Roche, a corporate color analyst, in the mid-1980s—the shift between colors associated with femininity and those influenced by men's fashion. She characterized the end of the 1970s as a dark phase in men's clothing with dusty[40] colors that were called elegant[41], refined[42], and sophisticated[43]. Use of these typical grayed European colors was new for American women's fashion but coincided with women's entry into fields such as finance, law, and medicine—fields formerly dominated by men. By the mid-1980s, those grayed colors looked dirty, and there was a shift to feminine colors. The shift between ultra-femininity and a suggestion of masculinity has been part of the fashion scene since Coco Chanel first introduced women to the concept of borrowing from men's closets[44] in the 1920s.

色彩周期与社会变化相关。一位企业色彩分析师 June roche 总结出一个明显的色彩周期，20 世纪 80 年代中期，色彩在女性化和男性化之间变化。她指出，20 世纪 70 年代后期，男性服装呈深色，尘土色被称为优雅、精致和成熟。美国女性时尚使用这些典型的欧洲灰色调，因为女性进入了一些一贯由男性主宰的行业，例如金融、法律和医学行业。到 20 世纪 80 年代中期，那些灰色调色彩看上去有些暗沉，又变化为女性化色彩。1920 年可可·香奈儿（Coco Chanel）将男性服装概念引用到女装上，从此，特别女性化到女装男性化的轮回变化就成为时尚舞台中的部分内容。

专业词汇

40. dusty adj. 浅灰色
41. elegant adj. 优雅的
42. refine vt. 精致
43. sophisticate adj. 成熟的
44. Closet n. 衣橱

Exercises

(1) Understanding the text.

Read the text and answer the following questions.

1) Do you think different colors of clothing bring different feelings effect to people?

2) Can you summarize the color cycles of fashion between 1999 and 2019?

3) Please describe Munsell Color System systematically?

(2) Building your language.

The following words and expressions can be used to talk about colors. Choose the right ones to fill in the blanks in the following sentences. Change the form where necessary.

survey	psychology	calm	luxury
stimulation	relaxation	associate	participant

If you're looking to cool your jets and relax, one of the most ____________ colors to be surrounded by, according to a global ____________ from paper merchant G. F. Smith and psychologists at the University of Sussex, is navy blue, closely followed by teal-like turquoise, and soft pastel pink.

The World's Favorite Color Project involved 26596 ____________ from over 100 different countries, to get some insights into the world's most beloved color.

"First, the more saturated the color is, the more it is associated with excitement and ____________," Professor Anna Franklin, a leading expert in color ____________ at the University of Sussex, explained. "Second, the lighter the color, the more it is associated with calmness and ____________. Many studies have found that blue and green are also associated with calmness and relaxation."

The findings also showed that orange is most often ____________ with happiness, while pink is viewed as the sexiest, and the colors people around the world most associated with ____________ are white, purple and orange.

(3) Sharing your ideas.

Please write a short introduction (of around 300 words) to explore the relationship between color and life cycle of products. Try to make full use of what you've learned from Text, including the relevant information from the reading text as well as words and expressions.

While everybody talks about fashion nowadays hardly anyone ever stops to consider what truly is what its origins are or how it came to occupy all spheres of society. In our consumerist culture, nothing escapes its influence, and it is therefore safe to say that today fashion has become a way of life. Conceived[2] in the context of dress fashion as a logic bases on novelty[3] has extended to all areas of society, a fact first confirmed in 1890 when the French Sociologist Gabriel de Tarde defined it as a social process independe[illegible] dress[illegible] are the only [illegible] tha[illegible] the [illegible] fashion cam[illegible]ing [illegible]ion are [illegible] ma[illegible] belong [illegible] as pointed [illegible] George Simm[illegible] easy because p[illegible] multiplied[5]. I[illegible] part of the cu[illegible]ture and th[illegible] can be s[illegible] from multip[illegible] angles [illegible] the p[illegible] of history sociology, anthropology psychology.

Unit 4 Fashion

第4单元 时尚

The Oxford English Dictionary defines fashion as "current popular custom or style, especially in dress".

While everybody talks about fashion nowadays, hardly anyone ever stops to consider what truly is, what its origins are, or how it came to occupy all spheres① of society. In our consumerist culture, nothing escapes its influence, and it is therefore safe to say that today fashion has become a way of life.

Conceived② in the context of dress, fashion as a logic bases on novelty③ has extended to all areas of society, a fact first confirmed in 1890, when the French Sociologist Gabriel de Tarde defined it as a social process independent of dress. Homo sapiens④ are the only animals that wear clothes, and fashion came into being because men and women are social animals who, while desiring to belong to a group, also want to be different, as pointed out by German sociologist George Simmel. Defining fashion is not easy, because fashion multifaceted⑤. It forms part of the culture and thus can be studied from multiple⑥ angles[1] from the perspective of history, sociology, anthropology, psychology, art, economics, or science. Fashion is a complex process that reflects society's transformations in each age.

Fashion has been influenced by wars, conquests, laws, religion, and the arts. Individual personalities have also had an impact on fashion. Fashion is now directly linked with film, music, literature, arts, sports and lifestyle as never before. The contribution of fashion and its growing influence has also permeated⑦ into other aspects of the business sector as has never before been witnessed. Ever fashion follows a cycle, and fashion cycle has no specific measurable time period. Some styles sustain⑧ for longer period or some die out soon and some styles come back years after it was declined⑨. So we can say fashion changes with time and has always been evolving to fit the taste, lifestyle and demands of society.

4.1 The origin and evolution of fashion

Fashion is a particular system of production[2] and organization of dress that emerged the West with the advent of modernity during the fourteenth century, subsequently⑩ expanding with the rise of mercantile⑪ capitalism, hand-in-hand[3] with technological processes. Pre-modern societies were traditional-based on worship of the past, of tradition-perpetuating the same forms of dress with negligible⑫ alterations⑬.

The system of fashion took root when a rupture⑭ from the past (from the old) in benefit of the future (the new) occurred, which is to say, when newness became a constant and general principle[4], highlighting a predilection⑮ typical of the West: modernity.

牛津英语词典定义时尚为“正在流行的风俗或样式，尤其指着装”。

今天虽然每个人都在谈论时尚，但几乎没有人停下来思考什么是真正的时尚，来源于何处，为什么占据社会的各个领域？在用户至上的文化中，什么事都逃不过时尚的影响，因此可以很有把握地说，今天时尚已经成为一种生活方式。

在着装背景下，时尚被认为以新奇为逻辑基础已经延伸到社会各个领域，这一事实在1890年首次得到证实，法国社会学家加布里埃尔·塔尔德把时尚定义为不受着装影响的社会过程。德国社会学家格奥尔格·齐美尔指出，所有动物中只有智人（现代人的学名，译者注）穿衣服。时尚的出现是因为男性和女性是社会性动物，他们既希望隶属于一个群体，又希望与其他人不同。定义时尚很难，因为时尚具有多面性。它是文化的一部分，因此可以从历史学、社会学、人类学、心理学艺术、经济学和科学等多个角度去研究它。时尚是一个复杂的过程，它反映了各个时期社会的变革。

时尚曾受到战争、征服、法律、宗教和艺术的影响。个人性格也对时尚产生影响。现在，时尚前所未有地直接与电影、音乐、文学、艺术、体育和生活方式联系在一起，也从未目睹时尚已经渗透到各行各业，并为之作出的贡献和日增的影响力。时尚永远遵循周期，时尚周期没有具体可测量的时间段。有些样式持续较长时期，有些很快消亡；而另一些时尚在它衰落以后又重新回归。因此，我们可以说时尚随时间的变化而变化，并一直不断发展，以适应品味、生活方式和社会需求。

4.1 时尚的起源和演变

时尚是服装生产和组织的特别体系，产生于14世纪西方现代化来临之际，随后在兴起的商业资本主义和技术进步过程中逐步发展。社会现代化之前，人们尊崇过去的传统，着装也延续着传统形式，其变化可以忽略不计。

时尚体系要与过去（旧的）彻底决裂，有利于未来（新的）产生。即当“新”成为一种持续的普遍原理时，就突显出西方一种典型的偏爱现代性。

专业词汇

1. angle n. 角度
2. production n. 产品
3. tradition n. 传统
4. principle n. 基本原理

通用词汇

① sphere n. 球
② conceive vt. 构思
③ novelty n. 新奇
④ Homo sapiens n. 智者
⑤ multifaceted adj. 多方面的
⑥ multiple adj. 多重的
⑦ permeate vt. 弥漫
⑧ sustain vt. 维持
⑨ decline vi. 下降
⑩ subsequently adv. 其后
⑪ mercantile adj. 贸易的
⑫ negligible adj. 可以忽略的
⑬ alteration n. 变化
⑭ rupture n. 断裂
⑮ predilection n. 偏爱

There is in fashion a vital trait of modernity: the abolition of traditions. Fashion has a characteristic of the modern because it is an indication of emancipation⑯ from, among other things, authorities. Fashion is irrational⑰. It consists of change for the sake of change whereas the self-image of modernity consisted in there being a change that led towards increasingly rational self-determination.

Modernization consists of a dual movement: emancipation always involves the introduction of a form of coercion⑱, since the opening of one form of self-realization always closes another. Every new fashion is a refusal to inherit⑲, subversion against the oppression of the preceding fashion. Seen in this way, emancipation lies in the new fashion, as one is liberated from the old one. The principle of fashion is to create an ever-increasing velocity⑳, to make an object superfluous as fast as possible, in order to let a new one have a chance.

Fashion is irrational in the sense that it seeks change for the sake of change, not in order to improve the object, for example by making it more functional[5]. It seeks superficial changes that in reality have no other assignment than to make the object superfluous on the basis of non-essential qualities, such as the number of buttons[6] on a suit jacket[7] or the famous skirt length[8]. Why do skirts become shorter? Because they have been long. Why do they become long? Because they have been short. The same applies to all other objects of fashion.

The evolution of the fashion system can be divided into three stages:

(1) Aristocratic㉑ fashion appeared in the second half of the fourteenth century and lasted until the middle of the nineteenth century. Its dominant figure[9] was masculine[10], with men exhibiting the full range of their power through a fashion based on ornamentation[11].

(2) Centennial fashion emerged in the second half of nineteenth century and extended up to the 1960s. Men were eclipsed by women, who drew attention to themselves with haute couture[12] design.

(3) Open fashion was born in the 1960s and continues to this day, characterized by the great interest of both sexes in their appearance, coinciding with the rise of consumer society.

4.2 Fads[13] and fashions

In academic and popular discussions, fads and fashions are often treated together.

They help fill in large culturally blank areas that haven't explained with other forms of collective obsessions㉒. Fads and fashions occur within nearly every sphere of social life in modern society, most obviously in the areas of clothing and personal adornment. The line separating fads and fashion is hardly clear, as both terms are frequently applied to aspects of change in the physical[14] presentation of the self and to areas not involving large economic investments.

时尚现代性的重要特征：废除传统。时尚具有现代性特征，是因为它表明了要从权威和其他事项中解放出来。时尚是非理性的，它为了变化而由变化构成，因为现代性的自我形象就是以变化而存在，是日趋理性的自我决定性的变化。

现代化由一种双重运动构成：解放总是涉及采用高压形式，因为一种自我实现形式的开始，总是要结束另一种形式。每一种新时尚就是拒绝继承和颠覆先前时尚的压迫。从这个角度看，解放在于新时尚，因为新的从旧的中解放出来。时尚的原理就是创造不断递增速度，使一种事物尽快地过剩，让新的获得机会。

时尚的非理性含义是为变化的缘故而寻求变化，而不是为了改进事物，例如不是使事物更具功能性。它试图改头换面，而事实上是改变非本质性特征，使事物成为多余，例如改变夹克套装上纽扣的数量或著名的裙摆长度。为什么裙子变短？因为它们一直是长的。为什么变长？因为它们一直是短的。这同样适用于时尚的其他所有对象。

时尚体系的演变分为三个阶段：

(1) 贵族时尚，产生于14世纪下半叶，持续到19世纪中叶。主导形象是男性化，以装饰为基础的男性时尚展示他们全方位的权力。

(2) 百年时尚，产生于19世纪下半叶，持续到20世纪60年代。男性的时尚地位让位于女性，男性将注意力转移到高级时尚设计。

(3) 开放性时尚，产生于20世纪60年代，并一直持续到今天。其特点是男女都对其外观有极大兴趣，与消费社会的兴起相吻合。

4.2 热潮和时尚

在学术和大众讨论中，热潮和时尚经常放在一起探讨。

它们很好地解释了集体着迷现象，此外没有其他形式能够解释这种现象，大大地填补了文化上的空白。在现代社会，几乎各个领域都有热潮和时尚的发生，着装和个性装饰最为明显。热潮和时尚之间没有明确的分界线，因为两个词汇经常用于个人身体外表的变化方面，涉及的范围没有大的经济投入。

专业词汇

5. function n. 功能
6. button n. 纽扣
7. jacket n. 夹克
8. length n. 衣长
9. figure n. 体型
10. masculine adj. 男性气质
11. ornamentation n. 装饰
12. haute couture n. 高级时装
13. fad n. 狂潮
14. physical adj. 身体的

通用词汇

⑯ emancipation n. 解放
⑰ irrational adj. 不合理的
⑱ coercion n. 强迫
⑲ inherit vi. 继承
⑳ velocity n. 速率
㉑ aristocratic adj. 贵族的
㉒ obsession n. 着魔

Fads and fashion can occur together and should be understood as expressive rather than instrumental㉓ forms of collective behavior. Because they are noninstrumental action they show a high degree of emotional involvement. Fads and fashions differ as well: fads are more spontaneous and tend to not follow the cycles that fashions do.

Fads are temporary, highly imitated outbreaks of mildly unconventional[15] behavior. In contrast a fashion is a somewhat long-lasting style of imitative behavior or appearance. A fashion reflects a tension between people's desires to be different and their desire to conform. Its very success undermines its attractiveness, so the eventual fate of all fashions is to become unfashionable. Fads also more frequently involve crowds and face to face interaction, whereas fashion usually involves diffuse[16] collectivity, in which widely dispersed individuals respond in a similar way to a common object of attention.

Fashion is much like fads and other collective obsessions, except that it are institutionalized and regularized, becoming continuous rather than sporadic㉔, and partially predictable[17]. Whereas fads often emerge from the lower echelons㉕ of society, and thus constitute a potential challenge to the class structure of society, fashion generally flows from the higher levels to the lower levels, providing a continuous verification of class differences. Continuous change is essential if the higher classes are to maintain their distinctiveness㉖ after copies of their clothing styles appear at lower levels. Thus, fads and fashions contribute to both social integration and social differentiation. With fashions tending to change cyclically within limits set by the stable culture. For fads and fashions established groups usually serve as the settings and conduits through which the behavior passes.

Fashion trends[18] change every season, bringing into trend diverse styles which can make you look fabulous. Fashion plays a very important role when it comes to the way others perceive us so paying as much attention to style can have a positive impact over the physical appearance. Fashion offers us new trends to choose from every season, allowing diversity to take over, avoiding this way a fashion routine.

4.3 Fashion cycle

The way fashion change is described as fashion cycle. Cycle means period of time or life span during which fashion exists.

热潮和时尚可以一起产生，应该理解为集体的表述行为而非工具性形式。因为它们是非工具性行为，显示出包含高度的情感成分。热潮和时尚也有区别：热潮具有自发性，不像时尚那样遵循周期。

热潮是短暂的、突发的极度模仿和温和的非常规行为。相比之下，时尚对行为和外貌的模仿持续的时间长一些。时尚反映了人们既希望与别人不同，又希望与别人趋同的矛盾。时尚非常成功地削弱自身的吸引力，因此所有时尚的最终命运是变得不时尚。热潮经常涉及大量人群和面对面的相互作用，而时尚通常涉及集体的传播，是广泛分散的个体关注一个共同的对象，以相似的方式做出反应。

时尚很像热潮和其他群体痴迷现象，但它被制度化和系统化，使其成为持续的而不是断续的，部分地可预测。然而，热潮经常在社会的下层出现，因此，潜在地构成了对社会等级结构的挑战。时尚一般是从高层流向低层，提供了阶级差别的连续验证。如果高层阶级的服装被低层阶级模仿后，高层阶级要维持他们的独特性那么不断变化就是本质。因此，热潮和时尚有利于社会的整合和社会差别化。稳定的文化设定了时尚只能在有限的范围内周期性变化。热潮和时尚通常是在一些既定的群体中，借助其环境和渠道，实现的行为。

时尚潮流每个季节在变化，带来了丰富多彩时尚样式，可以让你焕然一新。当提到别人看待我们的方式时，时尚起着十分重要的作用。因此高度注意风格，对我们生理外貌有正面影响。时尚为我们提供了多种新的潮流，供我们从每个季节中选择，使差异化产生，避免了循规蹈矩一种时尚路线。

4.3 时尚周期

时尚变化的方式可用时尚周期来描述。周期是指时尚存活的时间段或寿命。

专业词汇

15. unconventional adj. 非传统
16. diffuse adj. 传播的
17. predictable adj. 可预测
18. trend n. 趋势

通用词汇

㉓ instrumental adj. 起作用的
㉔ sporadic adj. 不定时发生的
㉕ echelon n. 等级
㉖ distinctive adj. 有特色的

4.3.1 The stages of the fashion cycle

The fashion cycle is usually depicted as a bell-shaped[19] cueve[20] encompassing five stages: introduction, increase in popularity, peak in popularity, decline in popularity, and rejection. Consumers are exposed very season to multitudes of new styles created by designers or launched[21] by big clothing brands[22]. It is seem some styles are rejected immediately by the buyers[23] on retail[24] level, where as some styles are accepted for a time as demonstrated by consumers purchasing and wearing them. With trend reports in new papers and fashion channels showing latest trends many women who consider themselves fashionable, or up to date with what's new, go out each season to assess what's needed in order to keep her wardrobe relevant. Then designers also are constantly going back in time for inspiration[25]. Each season a new version[26] of the old era is tapped and we see a few small changes to looks that have all walked down the catwalks[27] before.

4.3.1.1 Introduction of a style

Every designer each season works on a new collection[28], interpret their research into apparel. Every style has some different elements[29] like line[30], shape[31], colour[32], fabric. The first stage of the cycle where the new style is introduced may or may not be accepted by the consumers. Every style is reviewed at design centre and in fashion shows[33]. New styles are usually introduced in high price level. Usually a new style created by a designer is worn by the selected people who can afford it, and mostly these people are fashion leaders like celebrities[34] and rich people who loves to experiment and try out new styles to grab the attention of media. Such styles as they are expensive are produced in a small quantity.

4.3.1.2 Increase in popularity

A new style worn by a celebrity or famous personality, seen by may people and it may draw attention of buyers, the press, and the public. Most designers also have prêt-ɑ̀-porter[35] line that sells at comparatively low prices and can sell their designs in quantities. Manufacturers adopt design and styles to produce with less expensive fabric or les details[36]. The adaptations are made for mass production.

4.3.1.3 Peak in popularity

Styles at this stage are most popular. When production of any style is in volume[37], it requires mass acceptance. The manufacturers carefully study trends because the consumer will always prefer clothes that are in the main stream[38] of fashion. When a fashion is at height of its popularity, it may be in such demand that many manufacturers copy it or produce adaptations of it at many price levels. Length at this stage determines if the fashion becomes classic or fad.

4.3.1.4 Decline in popularity

A time comes after the mass production of a few styles people get tired and began looking for new styles. They still wear the particular style but are not willing to buy them at the same price. With the launch of new collection every season the popularity of the style of the previous seasons declines. Fashion is over saturated or flooded the market. Retail stores put such decline styles on sale rake as off season sale or clearing sale.

4.3.1 时尚周期的不同阶段

时尚周期通常用铃形曲线来描绘，包含 5 个阶段：引入、上升、高峰、下降和拒绝。每一个季节消费者可以看到很多新的款式，它们由设计师设计或由大的服装品牌公司推出。一些款式似乎在零售层面上立刻被购买者拒绝，而另一些款式被接受的时间长一些，它由消费者的购买和穿着来证明。那些认为自己很时髦的女性或认为紧跟新潮的女性，为了使她们的衣柜紧跟潮流，必须根据报纸报道的最新趋势，时尚频道播放的最新潮流，评估每个季节的新潮中自己需要什么。而设计师却要不断地回到过去寻找灵感。过去的款式往往被开发成新款，当在 T 台上展示时，我们看到了一些小的改动。

4.3.1.1 样式的引入阶段

每个设计师都要为每个季节的新产品工作，将他们的研究演绎成服装。每种样式有不同的元素，例如线条、款式、色彩和面料。在周期的第一阶段，一种新样式被引入时消费者可能接受或可能不接受。每一种样式在设计中心和时尚秀中被审查。新样式在引入阶段通常价格很高。设计师设计的新样式通常针对那些买得起的人，这些人大多是时尚领袖，如名人和富人，他们喜欢体验新样式，吸引媒体的注意。由于这些新样式的服装很贵，只能少量生产。

4.3.1.2 人数增长阶段

当很多人看到明星或名人穿新样式的服装时，可能会吸引购买者、新闻媒体和公众的注意。大多数设计师还有成衣生产线，以相对低的价格销售，可以批量销售他们的设计。生产商采纳设计和样式，用便宜的面料和较少的细节生产，这是为了适应大众产品。

4.3.1.3 人数高峰阶段

样式在这个阶段是最流行的。任何样式大批量生产时，需要大众接受。制造商仔细研究趋势，因为消费者总是喜欢服装处于时尚的主流阶段。当一种时尚处于流行的最高峰时，就会出现这样的情形，很多制造商复制它或者稍作改动，以不同的价格销售。在这一阶段时尚变成经典还是变成热潮，就决定这种时尚生命周期的长短。

4.3.1.4 人数下降阶段

一些样式在经过一段时间的大众产品阶段后，人们厌倦了，并且开始寻求新的样式。他们仍然穿着那些样式，但是不愿意再以相同的价格购买它们。由于每一季节有新的样式推出，前几季节流行的样式就衰落了。饱和或洪水般淹没的时尚市场就此结束。零售商店将这些过时的服装打折或清仓处理销售。

专业词汇

19. bel-shaped adj. 钟形
20. curve n. 弧线
21. launch vt. 发布
22. brand n. 品牌
23. buyer n. 买手
24. retail n. 零售
25. Inspiration n. 灵感
26. version n. 版本
27. catwalk n. T 型台
28. collection n. 系列时装
29. element n. 元素
30. line n. 线条
31. shape n. 形状
32. colour n. 色彩
33. show n. 发布会
34. celebrity n. 名人
35. prêt-à-porter 成衣
36. detail n. 细节
37. volume n. 体积
38. stream n. 潮流

4. 3. 1. 5 Dejection[27] period

It is the last phase of the cycle. Some consumers have already turned to new looks, thus beginning a new cycle. The rejection or discarding of a style just because it is out of fashion is called consumer obsolescence. Since consumers are no more interested manufactures stop producing the same and the retailers will not restock the same styles. Now it's time for a new cycle to begin.

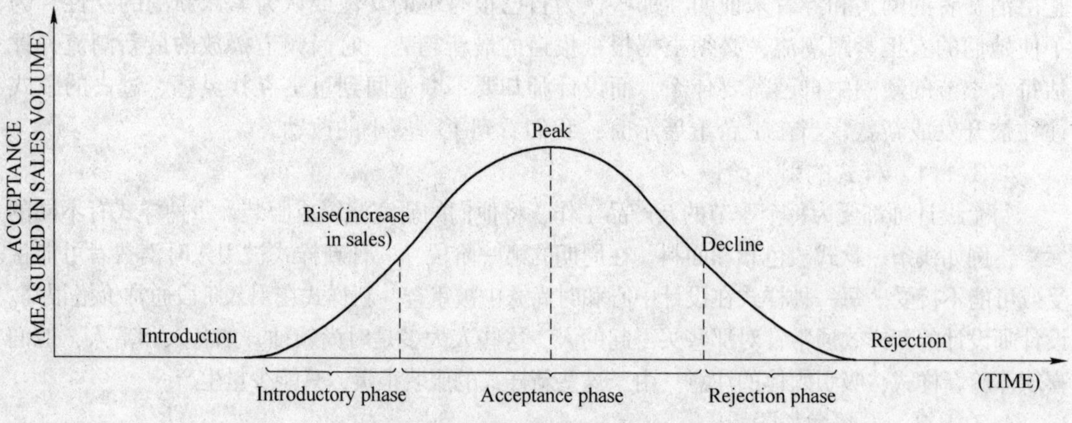

Figure 4. 1 Stages of a fashion cycle

4. 3. 2 Length of the cycles or fashion movement

Fashion cycle has no specific measurable time period. Ongoing motion of fashion in the fashion cycle is a movement. Fashion movement is affected by

(1) Economic or social factors;

(2) Invention of fibers or fabrics;

(3) Advertising of the product.

Rate of the movement varies with each fashion. Short time to peak in popularity, other takes longer; Some declines slowly, others swiftly.

4. 3. 2. 1 Classics

Fashion that always remains in the rise stage of the fashion cycle is known as classic.

The styles that remain more or less accepted for an extended period become completely obsolete. Example: Classic shirt, Jeans[39] Tailored suit.

4. 3. 2. 2 Fad

Fad also knows as short-lived fashion and can hold the attention of the consumer for a very short period. The consumer group is very small and the garments are low priced and flood the market in very short time. The consumer gets tired of the designs due to market saturation and they die out soon.

4.3.1.5 拒绝阶段

这是周期的最后阶段。一些消费者已经转移到新样式上，从而开始一个新的周期。一种样式的拒绝或丢弃，只是因为它过时了，被称为消费者过时。因为消费者不再感兴趣，制造商就停止生产这种样式，零售商也不再储存这种样式。现在是到了新周期开始的时候了。

4.3.2 周期长度或时尚运动

时尚周期没有具体的可测量的时间段。时尚周期中正在运行的动势是一种运动。时尚运动受以下因素的影响：

(1) 经济或社会的因素；

(2) 纤维或面料的创新；

(3) 产品的广告宣传。

每种时尚的运动速率不同，一些时尚在很短的时间内到达顶峰状态，而另一些时尚需要很长时间到达高峰；一些时尚在很长时间内衰落，而另一些在很短时间就衰落。

4.3.2.1 经典

时尚总是保持在周期的上升阶段，被称之为经典。

仍然有一些样式在长时间段里或多或少地被人们接受。这种样式永远不会变得完全过时。例如：经典衬衫、牛仔西装。

4.3.2.2 热潮

热潮是指短生命周期的时尚，在相当短的时间内吸引消费者的注意力。消费群体非常小，服装价格很低，在很短时间内充斥市场。由于市场饱和，消费者疲倦这种设计，因此它们死得很快。

专业词汇

39. jean n. 牛仔裤

通用词汇

㉗ dejection n. 忧郁

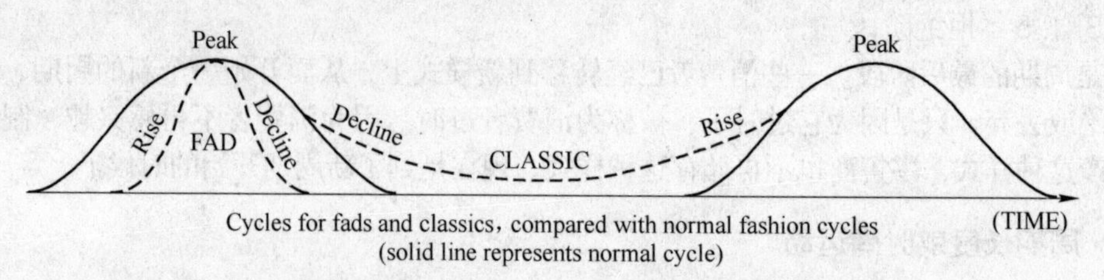

Figure 4.2 Cycles foe fads and classic, compared with normal fashion cycle

4.3.3 Recurring cycles

Fashion Designers draw inspirations from past. It has been noted that styles reappears years later and is reinterpreted for a new time. A change in element is normal like change in the silhouette or proportion may recur and is sometimes interpreted with a change in fabric and detail. Many nostalgic looks are drawn by designers from the 1940s, 50s, 60s, 70s, 80s. However, the use of different fabrics, colours, and details make the looks unique in every creation.

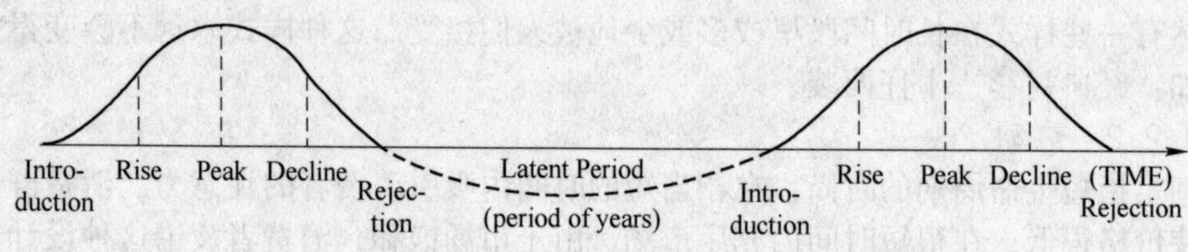

Figure 4.3 Recurring fashion cycle

4.3.3 周期复苏

时尚设计师从过去汲取灵感。已经注意到样式在多年后又再次出现，在新时期得到重新诠释。一般在轮廓或比例的要素方面作改变，有时通过面料和细节诠释变化。许多怀旧的样式就是设计师回归20世纪40年代、50年代、60年代、70年代和80年代样式，然而由于使用了不同的面料、色彩和细节，每一种设计呈现出独特的样貌。

Exercises

(1) Understanding the text.

Read the text and answer the following questions.

1) What is fashion?

2) Can you explain the modernization of fashion in China?

3) What are the differences between fashion and fads?

(2) Building your language.

Translate the following sentences into Chinese.

1) As the population ages and travel increases, people also want clothes that are comfortable to move in, sit in, travel in and so on.

____

2) Grading is the method used to increase or decrease the sample-size production pattern to make up a complete size range.

____

3) The federation includes three branches: couture, women's ready-to-wear, and men's wear.

____

4) The British also have an international reputation for classic apparel.

____

5) In 1991, more than half of all apparel sold in the United States was bought on sale.

____

(3) Sharing your ideas.

Please write a short introduction (of around 300 words) to explore the relationship between fashion and tradition. Try to make full use of what you've learned from Text, including the relevant information from the reading text as well as words and expressions.

Unit 5

The Types of Fashion

第 5 单元

时尚的类型

A one-of-a-kind piece or ensemble2 is the crème de la crème of fashion, and is fully customized and made to order3 for a specific client according to his or her exact measurements4 and specifications.② One-of-a-kind garments are considered the pinnacle [illegible] fashion world because only one of its kinds is in existence. Custom-made6 garments are crafted at the haute couture level, using only the f[illegible] fabrics, trims7, embroideries8, and appliqués9. The price point reflects that level, due to the quality of materia[illegible] used and the superior extent of detail and [illegible]orkmanship that goes into making cu[illegible] Custom clothing is often referred to [illegible] [illegible]sistance because it is considered [illegible] [illegible]tible③ showpiece at every level. It [illegible] by many to be an art form10. Fin[illegible] [illegible]om pieces are often displayed in muse[illegible] [illegible] around the world and sell for thousan[illegible] dollars at auction. A custom client may req[illegible] one piece or an entire wardrobe for a series [illegible] special events, such as black-tie11 galas. It is the responsibility of the

The fashion industry is divided into five main markets according to price point: Haute couture, designer, bridge, moderate① and mass. However, there are additional markets there are just as important to be aware of, including one-of-a-kind, bespoke[1], contemporary, secondary, private label, and discount. The following provide a listing and explanation of all of the fashion industry markets, from highest to lowest price point.

5.1 One-of-a-kind

A one-of-a-kind piece or ensemble[2] is the crème de la crème of fashion, and is fully customized and made-to-order[3] for a specific client according to his or her exact measurements[4] and specifications②. One-of-a-kind garments are considered the pinnacle of luxury[5] in the fashion world because only one of its kinds is in existence. Custom-made[6] garments are crafted at the haute couture level, using only the finest fabrics, trims[7] embroideries[8], and appliques[9]. The price point reflects that level, due to the quality of materials used and the superior extent of detail and workmanship that goes into making each piece. Custom clothing is often referred to as the pièce de resistance because it is considered a true, irresistible③ showpiece at every level. It is considered by many to be an art form[10]. Finished custom pieces are often displayed in museum exhibits around the world and sell for thousands of dollars at auction. A custom client may request one piece or an entire wardrobe for a series of special events, such as black-tie[11] galas. It is the responsibility of the designer to come up with each of those items according to a specified timeline and perhaps a personal branding theme[12] that the client wishes to be carried out throughout his or her customized ensemble.

Celebrities who are presenters at an awards show, or who have received industry award nominations, will often be seen wearing a custom dress designed especially for the occasion. Other custom clients may include a celebutante (a person who is famous for being famous), a jet setter or socialite who is attending an exclusive event. A debutante④ who is making her debut into society at the cotillion ball, a high-profile businesswoman who is being honored at a conference, a low-profile client who prefers to remain anonymous after receiving an inheritance, or anyone who has an appreciation for custom clothing.

Costume designers are fashion designers who design and create customized costumes for film, televise, performing arts and stage productions, fashion shows, special events or other performances for "talent" or show business personalities, actors, models, singers, dances, and other performers. The process sometimes involves extensive research of a historical component⑤, such as the replication⑥ of clothing from a particular era, needs to be reproduced. Once the research is complete, designs are sketched, and fabric is sourced and purchased, then draped on a form

时尚行业按照价位分为五个主要市场：高级时装、设计师品牌、过渡类服装、中等类服装和大众型服装。但是，还有其他一些重要市场不能忽视，包括一对一、定制、流行、次品牌、私有品牌和打折服装。下面从最高到最低价位提供了时尚行业所有类型市场的列表和解释。

5.1 一对一类型

一对一类型的服装或套装是一种唯一的服装，根据特定顾客，他或她的准确尺寸和规格，完全客户化定制。一对一的服装被看成是时尚界最奢侈的，因为其样式唯一，以高级时装水准制作，只用最好的面料、辅料、刺绣和珠绣。价位反映了它使用面料的质量、超细的细节和每一件所需要的手工。定制服装经常被看成是一件超级精品，因为在各个方面它都被看成是一件真正的、不可抗拒的展品，很多人认为是一种艺术品。定制服装的成品经常在世界博物馆里巡回展出，在拍卖会上能卖到几千美元。定制客户可能需要一件或一整套定制服装，去参加一系列特别活动，例如黑领带半正式礼服。设计师的责任是根据特定的时间日期完成每一件单品，也许客户定制服装里隐含客户希望的个性品牌理念。

在颁奖典礼上做主持的名人，或获得行业提名的人，经常看到他们在这种场合穿定制的服装。其他定制客户还包括那些频频出现在镜头前的名媛（出名是因为出身名门）、阔佬或上流社会人士，他们需要参加某个特别活动。初进社交界的上流社会年轻女子初次参加沙龙舞会，一个成功职业女性很荣幸参加某个会议，一个身份低的女性在得到了遗产后仍想保持匿名，或是任何想定制服装的那些客户。

戏剧服装设计师是设计和创造定制化服装的时尚设计师，主要为电影、电视、表演艺术和舞台产品、时装发布会、特殊事件设计服装；或其他表演"人才"或演艺界名人、演员、模特、歌手、舞蹈者，以及其他表演者。有时这个过程要对包含的历史因素进行广泛的研究，然后再现它们，例如对某个特定时期服装的复制。研究一旦完成，设计草图，寻找和购买面料，在人台上

专业词汇

1. bespoke adj. 订制
2. ensemble n. 全套服装
3. made-to-order 定做
4. measurement n. 度量
5. luxury n. 奢侈
6. custom-made 定制
7. trim n. 装饰
8. embroidery n. 刺绣
9. applique n. 缝饰
10. form n. 形式
11. black-tie 男子半正式礼服
12. theme n. 主题

通用词汇

① moderate adj. 中等的
② specification n. 规格
③ irresistible adj. 无法抗拒的
④ debutante n. 初进社交界上流社会年轻女子
⑤ component n. 成分
⑥ replication n. 复制

(i. e, mannequin[13]) or patterned and then produced. The costumes often require accessories, such as hats[14], headdresses[15], tiaras[16] and other jewelry, hosiery[17], masks[18], wigs[19], and footwear. The process may involve the creation of something unique, like a full-body cat suit for a musical. Costume designer John Napier won a Tony Award in 1983 for Best Costume Designer, for the Broadway musical Cats. A singer such as Britney Spears will need a completely customized wardrobe created for her worldwide concert tours, consisting of several head-to-toe outfits for each series of songs matching each corresponding stage set. So the costume designer will need to carry out a feeling in the costumes and ensembles that will correspond⑦ with the overall concert theme. Singer and actress Madonna had 85 costume changes in the movie Evita, which shows how important a role a costume designer plays in the overall production of a movie.

Certain unique circumstances come into place for a costume designer that an ordinary fashion designer would not necessarily encounter. For example, costume designers have to pay special attention to the needs of the particular person they are fitting. For a dancer, the fit of his or her clothing is critical in ensuring that movement is not inhibited during performance.

5.2 Haute couture

French for "high sewing", or "high fashion", haute couture (often referred to more informally as "couture"), describes handmade, made-to-measure[20] garments using only the most luxurious fabrics, such as the finest cashmere[21], fur[22], suede[23], leather, and silk sewn with extreme attention to detail by the most skilled seamstresses[24], often using hand-executed techniques. It is the fusion[25] of both costume and high fashion and is often seen on the most affluent and famous people.

The Chambre Syndicale de la Haute Couture is an association whose members include those companies that have been designated to operate as an haute couture atelier or house. Haute couture is a legally protected and controlled label and can only be used by those fashion houses they have been granted this designation by the French ministry of industry. This governing body annually reviews its membership base, which must comply with a strict level of regulations and standards in order to maintain membership. The membership list changes annually as a result of its stringent⑧ criteria.

(人体模型）进行设计或者画纸样，然后生产。这种服装经常需要配饰，例如帽子、头巾、冠状头饰和其他的首饰、袜子、面具和鞋类。这个过程也许包括一些独特的设计，例如为了《猫》音乐剧设计套装。戏剧服装设计师约翰·纳皮尔（John Napier）在1983年为百老汇音乐剧《猫》设计的服装获得了最佳戏剧服装设计师托尼（Tony）奖。例如歌手布兰妮·斯皮尔斯（Britney Spears）为世界巡回音乐会就需要设计一套定制的服装。从头到脚需要好几套服装，与每一首歌曲设置的舞台背景相吻合。因此，戏剧服装设计师需要在服装中夹带着情感，与整个音乐会主题相吻合。歌手和演员麦当娜（Madonna）在电影《贝隆夫人（Evita）》中服装更换了85次，这就显示出戏剧服装设计师在整个电影作品中起着极其重要的作用。

对戏剧服装设计师来说，需要再现某种特定场景，而对于服装设计师来说，就不会碰到这个问题。例如，戏剧服装设计师必须考虑设计服装适合特定的人物一个舞蹈者，他或她的服装在表演时要确保能够活动自如，而不能受到压制。

5.2 高级女式时装

"Haute Couture"法语意思为高级缝制或高级时尚（不正式使用时直接称"Couture"），比喻手工制作、量体裁制的服装，仅使用最昂贵的面料，例如上等羊毛面料、裘皮、小山羊皮、皮革和丝绸，由最熟练的缝纫师精工细作，经常使用手工技术。它是戏剧服装和高级时尚融合，经常看到最富有和最有名的人穿着。

法国时装协会由已经获得高级时装工作室或时装屋的资格成员构成。高级时装是受法律保护和受控制的品牌，只有那些被法国行业部授权了的公司才能使用。这个管理机构每年审查它的成员基地，基地必须严格遵守规章和标准，以便维持会员资格。由于标准严格，成员名单每年都在变化。

专业词汇

13. mannequin n. 人体模型
14. hat n. 帽子
15. headdress n. 头饰
16. tiara n. 冠状头饰
17. hosiery n. 袜类
18. mask n. 面具
19. wig n. 假发
20. made-to-measure 量体裁制
21. cashmere n. 山羊绒
22. fur n. 毛皮
23. suede n. 小山羊皮
24. seamstress n. 女裁缝
25. fusion n. 融化

通用词汇

⑦ correspond vi. 相应
⑧ stringent adj. 严格的

The couture house is headed by a fashion couturier who oversees a workroom of skilled worker who practice their hand-made craft[26] as experts in wither dressmaking or tailoring. The process may begin with a sketch, an illustration, or a draped and cut muslin[27] or toile[28], depending on the designer's preference. To finalize a couture piece, fine trim embroidery, and embellishments[29] are often purchased by outside sources, who are expert practitioners⑨ in their respective field and then meticulously sewn into each piece. Exquisite fit is an inherent⑩ quality of couture piece. The client will endure a series of fittings to determine that exact measurements have been achieved, to ensure not only precise fit but also style and comfort, which are equally essential.

When haute couture collections were first produced, they were presented to the press, buyers, and high-end clientele in a trunk show[30] format in a designated salon. Each model carried a card that indicated a corresponding look number, making it easy for those in attendance to jot down the garments that were to their liking. Once selections were made, the client would sit with the designer, who would then fit the garments to that client's specific measurements and exact preferences, or a buyer would reproduce them for their own store.

Today, the couture collections are seen on the runways during pairs Fashion Week. Pricing typically begins in the high thousands and can reach into the hundreds of thousands for these fine garments. Many companies use the glamour[31] and appeal of their couture collections, which account for a small market share of their overall business, as a catalyst to boost sales for their ready-to-wear, accessories, and fragrance[32] business, which represent the bule of their revenue. Couture collections are often used as a "visual advertisement" to bring excitement to the brand and to elicit⑪ sales for the more affordable ready-to-wear collection Such as some well-known couture labels are Armani Privè, Atelier Versac, Chanel, Christian Dior, Givenchy, Jean Paul Gaultier, and Valentino.

The Chambre Syndicale de la Haute Couture accepts "foreign" members; however, there are only a handful of fashion designers outside of Paris who practice the fine technique of couture craftsmanship. Elie Saab, Giorgio Armani, and Paul Smith are examples. The French Ministry allows for outside members in an effort to show their strong belief in the importance of the globalization of the fashion industry. Ralph Rucci, Rick Owens, Adam Kimmel, Zac Posen, and Mainbocher are the only American designers to have achieved haute couture status.

They have each been invited by the Ministry to show their collections in Paris and currently are, or have been, members of *The Chambre Syndicale de la Haute Couture*. Interestingly enough, Thom Browne, a New York-based menswear designer, independently showed his collection in Paris as a nonmember. A complete list of current members can be found at www. modeaparis. com.

高级时尚屋由高级女裁缝师或裁缝师监管工作室里所有熟练工人从事手工工艺。根据设计师的偏好，这个过程也许从草图或效果图开始，或从立体裁剪和剪裁白坯布开始。为了完成一件高级时装，精致装饰、刺绣和修饰经常是从外面购买，有人专门从事这个行业，然后谨小慎微地缝合到服装上。高级时装必须精确试衣。顾客要忍受很多次试衣，决定精确的尺寸，确保不仅精确地合身，还要有样式和舒适，这些都相当重要。

最新生产的高级时装系列，要在指定的展厅里以非公开的形式展示给新闻媒体、买手和高端客户。每一个模特儿拿着一张与样式编号对应的卡片，使观众能够草草地记下他们喜欢的服装。客户将选中的款式告知设计师，设计师将按照客户的实际尺寸和准确参数制作合身的衣服，或者买手将为自己的商店要求重新生产他们选中的款式。

今天，在巴黎时尚周上的T台上能够看到高级时装的发布会。这些上等的服装的价格从最低几千美元，到高达几万美元。很多公司利用高级时装发布会的优雅和魅力，作为催化剂，促进他们的成衣、配饰和香水生意的销售，因为这些代表了他们全年的收益率，而高级时装只占他们整个生意的一小部分。高级时装发布会经常被当作视觉广告，给品牌带来活力，给相对便宜些的成衣带来更多的销售。例如一些知名的高级时装品牌：阿玛尼（Armani Prive），范思哲（Atelier Versace），香奈儿（Chanel），克里斯汀·迪奥（Christian Dior），纪梵希（Givenchy），让·保罗戈尔捷（Jean Paul Gaultier）和华伦天奴（Valentino）。

法国高级时装协会接纳外国成员，但是接受巴黎以外的时尚设计师很少，他们需要具备娴熟的高级时装技术，例如艾利·萨伯（Elie Saab），乔治·阿玛尼（Giorgio Armani）和保罗·史密斯（Paul Smith）。法国政府允许外来成员在这样一个全球化时尚行业占有重要地位的国度里施展他们的才能。拉尔夫·鲁奇（Ralp Rucci），瑞克·欧文斯（Rick Owens），亚当·基梅尔（Adam Kimmel），扎克珀森（Zac Posen）和梅因布彻（Mainbocher）是取得这种资格的美国设计师。

他们被法国政府邀请到巴黎举行他们的发布会，他们已经是巴黎高级时装协会的成员。非常有趣的是，男装设计师桑姆·布朗尼（Thom Browne），他的总部位于纽约，没有协会成员资格，却在巴黎独立举行了他的发布会。在 www. modeaparis. com 网站上能够看到现时会员完整列表。

专业词汇

26. craft n. 技艺
27. muslin n. 平纹细布
28. toile n. 坯布
29. embellishment n. 修饰
30. trunk show 非公开时装
31. glamour n. 魅力
32. fragrance n. 香水

通用词汇

⑨ practitioner n. 从业者
⑩ inherent adj. 固有的
⑪ elicit vt. 引出

5.3 Bespoke

Bespoke is a British term used to describe individually crafted and patterned men's clothing. *The Oxford English Dictionary* defines "bespoke" as made-to-order clothing, made to each individual customer's precise measurements and specifications. Although bespoke is not a protected label, like couture, *The Savile Row Bespoke Association* (a professional organization consisting of Savile Row tailors) has attempted to set a standard by providing minimum requirements for a garment to be allowed the prestigious⑫ use of its name. Savile Row is a very short street in central London, called "the golden mile of tailoring", famous for its bespoke tailors, among them Davies and Son, Gieves & Hawkes, and Norton and Sons Historical Savile Row clients have included Napoleon Ⅲ and Winston Churchill.

5.4 Designer

Also known as ready-to-wear (often abbreviated RTW) or "off-the-rack" and by the French term Prêt-ɑ̀-porter, designer is factory made and finished to fit standard sizes. Don't, however, let the phrase "off-the-rack" fool you. Whether mass produced or offered in limited quantities, designer clothing is exclusive and uses the finest imported fabrics and trims. Ready-to-wear collections are general presented twice a year (Spring/Summer and Autumn/Winter) during fashion weeks around the world, and they appear earlier than the couture collections. The price point can often exceed $1000 per garment, but can range in lower price points or skyrocket to high three-figure numbers. Some of the most popular designer labels are Ralph Lauren, Donna Karan, Vera Wang, and Catherine Malandrino.

In Pairs, *The Chambre Sydicale du Prêt-à-porter des Couturiers et des Greateurs de Mode* is an established association, created in 1973, that is made up of all the fashion designers who produce ready-to-wear. *The Chambre Syndicale de la Mode Masculine* is an association that specifically includes the top menswear designers who produce ready-to-wear collections.

5.5 Bridge

Bridge garments are in between ready-to-wear and better, and carry a price point generally ranging in price from $300 to $600 per garment. Career wear and separates[33] along with dresses, are often indicative of a bridge classification. DKNY, CK, and Anne Klein Ⅱ are examples of bridge labels.

5.6 Better

Better is one step down from bridge. Sportswear, various coordinates⑬ separates, and dresses may all appear in better collection, and will typically sell for less than $600 per piece, but they primarily fall into a price point range of $150-$300. Some of the most popular better labels are Ellen Tracy, Kenneth Cole, and Anne Klein.

5.3 定制

Bespoke（定制）是英国用语，用来描述男性服装个性化的工艺和样式。《牛津英语词典》定义“bespoke”为根据每个客户的精确尺寸和详细规格定做的服装。尽管定制不像高级时装是受保护的品牌，但是萨维尔街定制协会（由萨维尔街裁缝构成的专业组织）已经设法建立一套标准，达到最低标准的服装就被允许有萨维尔街的荣誉使用权。萨维尔街在伦敦中心，是一条很短的街，被称为“裁缝业的黄金一百米”，以定制裁缝而出名，他们当中有戴维斯父子（Davies & Son），君皇仕/吉凡克斯（Gieves & Hawkes）和诺顿父子（Norton & Sons）。历史上，拿破仑三世和丘吉尔都是萨维尔街的客户。

5.4 设计师品牌

设计师品牌也称为成衣（经常缩写为 RTW）或者“现成的”，法语称 Pret-a-porter，设计师服装是根据标准尺寸由工厂制造完成。但是，你不要被短语“off-the-rack”所愚弄，不管是大规模或有限数量生产，设计师服装是高级的，使用最好的进口面料和装饰。世界各地的成衣发布会通常在时尚周期间举行，一年两次（春/夏和秋/冬），它们比高级时尚发布会早。通常每件价格超过 1000 美元，但范围可以从较低价位或扶摇直上到三位数。最有名的一些设计师品牌是拉尔夫·劳伦（Ralph Lauren）、唐娜·凯伦（Donna Karan）、王薇薇（Vera Wang）和凯瑟琳·玛兰蒂诺（Catherine Malandrino）。

在巴黎，“裁缝和时尚设计师”是一个协会，建立于 1973 年，由所有生产成衣的时尚设计师构成。“裁缝和时尚男性设计师”是专门生产男装成衣的设计师组织。

5.5 过渡类型服装

过渡类型服装处于成衣和较好类型服装之间，每件服装的价格带通常在 300 ~ 600 美元之间。职业服装、可配套穿的单件服装和连衣裙等通常归入此类。唐可娜儿（DKNY），卡尔文·克莱因（CK）和安妮·克莱因Ⅱ（Anne Klein Ⅱ）就是过渡性服装品牌。

5.6 较好类型服装

较好类型服装比过渡类型服装低一等。运动型服装、各种可搭配穿的服装、单件服装、连衣裙，都可归入此类，每件服装通常不会超过 600 美元，价格带通常在 150 ~ 300 美元之间。最为人们熟知的“较好”品牌是爱伦·瑞丝（Ellen Tracy）、凯尼斯·柯尔（Kenneth Cole）和安妮·克莱因（Anne Klein）。

专业词汇

33. separates n. 可搭配穿着的女服

通用词汇

⑫ prestigious adj. 受尊敬的　　⑬ coordinate n. （女性）配套衣物

5. 7 Contemporary

Contemporary collections offer trendy apparel at a relatively affordable price point aimed at women in their twenties and thirties. Cynthia Steffe, Rebecca Taylor, and BCBGMAXAZRIA are all considered contemporary designers.

5. 8 Secondary

Secondary lines are used by designers who want to offer a lower-priced line aside from their designer collection. The price points differ, but these fashions can generally be found for less than $300 per piece at retail. Marc by Marc Jacobs and Lauren by Ralph Lauren are considered secondary lines.

5. 9 Moderate

Moderate fashions are promoted to the average, everyday customer and usually retail for less than $100 a piece. Some of the most popular moderate retailers are Liz Claiborne, Abercrombie & Fitch, Nine West, and the Gap.

5. 10 Private label

Merchandise[34] that is manufactured by a store, or in partnership with an apparel manufacturer, is considered private label. Store advantages include greater control over production, cost[35], pricing, advertising budget[36], and design. Private label runs a gamut of price points and is generally produced for the bridge to moderate markets. Some of the most successful private label businesses are International Concepts (I. N. C) for Macy's and Hunt Club for J. C. Penney.

5. 11 Mass

Mass market or budget caters to the lower end of the apparel continuum, with retail pricing generally under the $50 price point. Product categories[37] generally include casual sportswear such as T-shirts[38] and jeans. Some of the most popular budget retailers are Old Navy, Target, Wal-Mart, Kmart, and Kohl's. Mass market is made in large quantities and is geared toward the general public.

5. 12 Discount

Discount merchandise, also referred to as off-price, is excess merchandise that not sell at its full retail price through its original and intended retailer. These items can be found at varying price points in an array of retail outlets such as Filene's Basement (the inventor of the off-price store concept), Ross Stores, T. J. Maxx, Loehmann's, Marshalls, and Saks Fifth Avenue. Discount merchandise can also be found in factory outlet stores.

5.7 现代类型服装

现代类型服装是指潮流性服装，价格相对便宜，目标顾客为 20 ~ 30 岁。辛西娅·史黛菲（Cynthia Steffe）、瑞贝卡·泰勒（Rebecca Taylor）和 BCBGMAXAZRIA 都被认为是现代类型服装的设计师。

5.8 二线品牌

二线品牌是指设计师除了他们的设计师品牌以外，还有一个价位低一些的品牌。不同二线品牌价位不同，但通常每件零售价格不超过 300 美元。马克·雅可布之马克（Marc by Marc Jacobs）和拉尔夫·劳伦之劳伦（Lauren by Ralph Lauren）就是二线品牌。

5.9 适中类型服装

适中类型服装是针对一般日常的消费者，通常每件价格不超过 100 美元。最大众的适中类型品牌有丽兹·克莱本（Liz Claiborne），阿贝克隆比 & 费奇（Abercrombie & Fitch），玖熙（Nine West）和盖璞（Gap）。

5.10 自有品牌

服装商品由服装商店或合作伙伴关系的服装生产商生产，这种品牌被认为是自有品牌。商店的优势是能够较好地控制产品、成本、定价、广告预算和设计。自有品牌价位变化性很大，通常从过渡类型服装到适中类型服装。最成功的私有品牌有美国顶级连锁百货公司梅西百货（Macy's）的自有品牌 International Concepts（I. N. C.）和 J. C. Penney 百货公司的 Hunt Club 自有品牌。

5.11 大众市场服装

大众市场或价格低廉的市场，满足较低端的消费者。每件零售价格在 50 美元以下。产品种类通常包括休闲运动服装，例如 T 恤和牛仔。最熟知的大众市场零售商有老海军（Old Navy），塔基特（Target），沃尔玛（Wal-Mart），凯马特（Kmart）和科尔士（Kohl's）。大众市场服装生产量大，针对普通大众。

5.12 打折服装

打折商品即降价商品，指按照零售商的定价未能卖出的剩余产品。这些商品在很多出清的零售商店都有销售，例如菲妮斯地下商场（Filene's Basement，降价商店概念的发明人），罗斯服饰零售店（Ross Stores），T. J. Maxx，Loehmann's，Marshalls 和第五大道（Saks Fifth Avenue）。打折商品也能在工厂的零售商店里买到。

专业词汇

34. merchandise n. 商品
35. cost n. 成本
36. budget n. 预算
37. category n. 类型
38. T-shirt n. T 恤

Within these price points, clothing classifications fall into various product categories, including women's, men's, young men's, collegiate, tweens (pre-teen), juniors, children's and layette[39] wear, formalwear[40], outwear, intimates⑭, maternity, and swimwear.

在这些价位之内，服装有很多种类，包括女装、男装、年轻男装、学院装、青年装、少年装、童装和婴幼儿装、正装、外出装、内衣、孕妇装和泳装等。

专业词汇

39. layette n. 婴儿的全套服装　　40. formalwear n. 正式服装

通用词汇

⑭ intimate n. 亲密

Exercises

(1) Understanding the text.

Read the text and answer the following questions.

1) By understanding the different categories of fashion, what are you going for?

2) Can you list more fashion designers?

3) Please introduce *The Chambre Syndicale de la Haute Couture* specifically.

(2) Building your language.

The following words and expressions can be used to talk about fashion. Choose the right ones to fill in the blanks in the following sentences. Change the form where necessary.

ensure	style	outfit	fashion
comfort	category	occasion	available

Fashion styles that are ____________ for women can keep them looking their best and can keep the heads turning with the way that the clothes are able to bring out certain looks in women. If you want to make sure that you look your best and can find ____________ at the same time, than you will want to make sure that you have possible fashion available to help you to get the ____________ that you want. The different types of women's fashion that are available will ____________ that you are able to continue looking your best and presenting yourself in a way that fits your style the best.

When you begin to look into women's fashion, you will notice that there are several ____________ that are available in order to help you to look your best. Knowing the differences between these and the types of ____________ that are included in each style will allow you to make the best determination about what will fit you best. The result will be the ability to have different styles available for every ____________ and for every look. Starting your search for the right ____________ by knowing the differences between these categories will ensure that you are able to prepare for every situation through the styles.

(3) Sharing your ideas.

Please write a short introduction (of around 300 words) to explore the types of Men's Fashion. Try to make full use of what you've learned from text, including the relevant information from the reading text as well as words and expressions.

For fashion designer, it is important to develop an awareness of your taste and style (not how you dress, designer are often the worst dressed in a room because they are to busy thinking about how to dress others). Not everyone has an aptitude ① or desire to design unconventional clothes. Some designers focus on the understatements or detail of garments. Other designers design conventional garments but it is the way they are put together that makes the original modern.

Knowing w[illegible] you are [illegible] that is essential but doesn't mean [illegible]u sh[illegible] not experiment. It can take [illegible] [illegible]self and this period of di[illegible] at college. There ha[illegible] [illegible]nt of soul searching [illegible] [illegible]ing the designer that you wa[illegible] [illegible]her finding out the designer t[illegible] [illegible]u must be true to your own [illegible] ho[illegible] you want to dress somebody.

Beyond that, the rest is in the hands of the industry and the fashion buying public to decide

Unit 6

Fashion Design

第 6 单元

时尚设计

For fashion designer, it is important to develop an awareness of your taste and style (not how you dress-designer are often the worst dressed in a room because they are to busy thinking about how to dress others). Not everyone has an aptitude① or desire to design "unconventional" clothes. Some designers focus on the understatements or detail of garments. Other designers design "conventional" garments, but it is the way they are put together that makes the original modern.

Knowing what you are best at is essential, but doesn't mean that you should not experiment. It can take a while to "know yourself" and this period of discovery is usually spent at college. There has to be a certain amount of soul searching; it's not so much being the designer that you want to be, but rather finding out the designer that you are. You must be true to your own vision of how you want to dress somebody.

Beyond that, the rest is in the hands of the industry and the fashion-buying public to decide, and for every person who likes your work there will be someone who doesn't. This is common and working in such a subjective② field can be confusing, but eventually you will learn to navigate your way through criticism and either develop a steely exterior③ or recognize which opinions you respect and which to disregard④. Once you accept this, you are free to get on with what you are best at-designing clothes.

6.1 Know your subject

If a career in fashion is what you want then you need to know your subject. This might appear to be an obvious statement, but it must be said. You may protest, "I don't want to be influenced by other designers' work". Of course not, but unless you know what has preceded⑤ you, how do you know that you aren't naively reproducing someone else work?

Magazines are a good place to start, but don't just automatically⑥ reach for Elle and Vogue. There are many more magazines out there, each appealing to a different niche market and style subculture⑦ and you should have a knowledge of as many as possible; they are all part of the fashion machine. Magazines will not only make you aware of different designers, but so-called life-style magazines will also make you aware of other design industries and culture events that often influence (or will be influenced by) fashion.

By regularly reading magazines you will also become aware of stylists, journalists, fashion photographers and hair and make-up artists, models, muses[1], brands and shops that are all-important to the success of a fashion designer. There are also some great websites that show images of outfits on the catwalk almost as soon as the show has taken place.

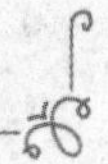

对时尚设计师来说，重要的是，要有发展自己品味和风格的意识（不是你穿着如何，在一个房间里，设计师经常穿得最糟糕，因为他们总是忙于想着如何打扮别人）。不是所有人都能够或希望设计非传统的服装。一些设计师专注于服装的陈述或细节；一些设计师设计传统服装，也正是这样，他们共同制造了新颖的现代时尚。

知道你最擅长什么是重要的，但并不意味着你可以不去尝试。认识自己需要段时间，通常是在大学的这段时间发生，那是一种灵魂的反省。与其说你想成为一名设计师，不如说你想成为什么样的设计师。你必须真实地用自己的视角考虑如何去打扮别人。

除此之外，其他方面就是由行业和大众时尚购买者决定，因为有人喜欢你的作品，有人不喜欢你的作品。在这种主观性行业里工作，面临这样的困惑是很正常的。但是，通过批评最终你将学会驾驭你自己，既能锻炼自己钢铁般的内心，而且能够识别哪些意见值得你尊重，哪些意见值得放弃。一旦你接受这些，你便易于了解你最擅长设计什么服装。

6.1 熟悉你的领域

如果你想要从事时尚职业，你需要熟悉你的领域。这似乎是显而易见的，但必须说。你也许声称“我不想受到其他设计师作品的影响”。当然不，除非你知道你的前人做了些什么，否则，你怎么知道你不是在天真地复制别人的作品。

杂志是起步的好地方，但不能想当然地就只看 Elle 和 Vogue 两本杂志。还有很多杂志，每一种杂志，关注不同的个性市场和亚文化风格，你必须尽可能地了解更多的知识，它们都是时尚机器的零部件。杂志不仅使你了解不同的设计师，同时，生活方式方面的杂志还将使你了解到其他的设计行业和影响时尚的文化事件（或受到时尚影响的文化事件）。

通过定期阅读杂志，你将对造型师、新闻记者、时尚摄影师、发型师、化妆师、模特儿、缪斯、品牌和商店有更多的了解，所有这些对时尚设计师的成功是很重要的。另外，还有很多网站张贴了 T 台上展示的服装，它们发布的速度几乎与 T 台同步。

专业词汇

1. muse n. 缪斯

通用词汇

① aptitude n. 才能
② subjective adj. 主观的
③ exterior n. 外部
④ disregard vi. 不顾
⑤ precede vt. 先于
⑥ automatically adv. 自动地
⑦ subculture n. 亚文化

6.2 Starting your research

Designers are like magpies⑧, always on the lookout for something to use or steal! Fashion moves incredibly⑨ fast compared to other creative industries and it can feel like there is constant pressure to reinvent the wheel each season. Designers need to be continually seeking new inspiration in order to keep their work fresh, contemporary, and above all, to keep themselves stimulated ⑩.

In this sense, research means creative investigation, and good design can't happen without some form of research. It feeds the imagination and inspires the creative mind.

Research takes two forms. The first kind is sourcing material and practical elements. Many fledgling⑪ designers forget that finding fabrics and other ingredients⑫-rivets[2], fastenings[3] or fabric treatments[4], for example—must make up part of the process of research and having an appreciation⑬ of what is available, where from, and for how much, is essential.

The second form of research is the kind you make once you've found a theme or concept for use in your designs. Themes can be personal, abstract or more literal. Alexander McQueen, Vivienne Westwood and John Galliano have designed collections where the sources of inspiration are clear for anyone to see. McQueen's *it's A Jungle Out There 1997—1998* collection mixed religious paining with the evocation⑭ of an endangered⑮ African antelope Westwood has drawn on pirates, the paintings of Fragonard and 17th-and 18th-century decorative arts in the Wallace Collection for inspiration in different collections. Galliano has been influenced by the circus, ancient Egypt, punk singer Siouxsie Sioux and the French Revolution.

Designers may also convey a mood[5] or use a muse for inspiration. Galliano currently cites singer Gwen Stefani as a muse, but has also based collections around 1920s' dancer Josephine Baker and Napoleon's Empress Josephine.

Using a theme or concept makes sense because it will hold together the body of work, giving it continuity and coherence[6]. It also sets certain boundaries[7]—which designer is free to break—but having a theme initially⑯ gives the designer focus.

6.3 Choosing a concept

When choosing a theme, be honest, it needs to be something that you can work and live with for the duration⑰ of the collection. This means that it should be a subject that you are interested in, that stimulates you and that you understand.

6.2 开始你的研究

设计师就像喜鹊那样，总是蹲守在瞭望台上，寻找可以使用或偷的东西。与其他设计行业相比，时尚的变化速度令人难以置信，你感觉到持续的压力，每一个季节都要重新发明转动的车轮。为了保持作品的新鲜度和时代感，设计师需要持续寻找新的灵感。总之，要使自己处于兴奋状态。

在这种意义上来说，研究意味着创造性调查，如果没有某些形式的研究，就没有好的设计产生。调查促进想象，激发创造性的智慧。

研究有两种形式。第一种是寻找资料和实际元素。很多没有经验的设计师，往往忽略寻找面料和其他元素，例如，铆钉、松紧带、面料的处理，它们都属于研究的过程。做评估是必要的：什么是可以得到的，它们来自哪里，需要多少？

第二种研究形式是在你的设计中寻找一种主题或概念。主题可以是具象的抽象的或者更加理论化的。亚历山大·麦昆（Alexander McQueen），维维安·韦斯特伍德（Vivienne Westwood）和约翰·加利亚诺（John Galliano），任何人都可以清楚地看到他们设计作品的灵感源。麦昆 1997—1998 年的“丛林之中”系列融合了宗教绘画以及对非洲羚羊濒临灭绝的呼唤。韦斯特伍德的海盗系列，以华莱士收藏馆（Wallace Collection）里弗拉戈纳尔（Fragonard）的绘画和 17 世纪、18 世纪的装饰艺术为灵感。加利亚诺（Galliano）受到了马戏、古埃及、朋克歌手苏克西（Siouxsie Sioux）和法国大革命的影响。

设计师还可以传达一种情绪，或者以一位缪斯为灵感。加利亚诺以歌手格温·史蒂芬妮（Gwen Stefani）作为他的缪斯，在他的系列中还以 20 世纪 20 年代的舞蹈家约瑟芬·贝克（Josephine Baker）和拿破仑皇后约瑟芬（Josephine）为灵感。

使用一个主题或概念将会找到一种感觉，能够把握作品的主题，具有连贯性和一致性。它也会设定一种界限，当然，这种界限设计师可以轻易地打破，但是，有了主题设计师可以更加集中注意力。

6.3 选择一种概念

诚实地选择一个你能为之工作和持久生活的主题，也就是说你感兴趣的、你为之激动和了解的主题。

专业词汇

2. rivet n. 铆钉
3. fastening n. 系扣
4. fabric treatment 面料后处理
5. mood n. 情绪
6. coherence n. 统一
7. boundary n. 边线

通用词汇

⑧ magpie n. 喜鹊
⑨ incredibly adv. 不可思议地
⑩ stimulate vt. 刺激
⑪ fledgling adj. 刚开始的
⑫ ingredient n. 组成部分
⑬ appreciation n. 欣赏
⑭ evocation n. 唤出
⑮ endanger vt. 使遭受危险
⑯ initially adv. 开始
⑰ duration n. 持续

Some designers prefer to work with an abstract[8] concept that they want to express through the clothing (for example, "isolation"), while others want to use something more visually orientated (such as "the circus").

Either of these approaches is appropriate and it is about choosing which works for you. But it does need to work for you; it is pointless choosing a theme that doesn't inspire you. If the ideas are still struggling to come after a certain point a clever designer will be honest and question their choice of theme.

Remember, press and buyers are generally only interested in the outcome. Do the clothes look good? Do they flatter? Do they excite? Will they sell? They are not necessarily interested how well you've managed to express quantum[18] physics through a jacket. But if this is what you want to express, then do it.

6.4 Research and the sources

Where to go begin your research depends on your theme or concept. For an enquiring designer the act of researching is like detective work, hunting down elusive information and subject material that will ignite[19] a spark.

The easiest place to start to research is on the internet. The Web is a fantastic source of images and information. It is also greet for sourcing fabrics direct from manufactures that produce socialist material or companies that perform specific service.

A good library is a treasure. Libraries are geared to provide books to a broad cross section of the community so tend to have a few books about many subjects. Specialist libraries are the most rewarding, and the older the library the better—books that are long out of print will (hopefully) still be on the shelves, or at least viewable upon request. Colleges and universities should have a library geared towards the courses that are being taught though access may be restricted if you are not actually studying there.

Flea markets and antique[20] fair are useful sources of inspirational objects and materials for designers. It goes without saying that clothing of any kind, be it antique or contemporary, can inspire more clothes. Historic, ethnic[21] or specialist clothing—military garments, for example—offer insight into details, methods of manufacture and construction that you may not have encountered before.

Like flea markets, charity shops are great places to find clothes, books, records and bric-a-brac that, in the right hands and with a little imagination, could prove inspirational. Everyday objects that are no longer popular or are perceived as kitsch[22] can be appropriated rediscovered and used ironically to design clothes.

Museums, such as London's Victoria and Albert Museum, not only collect and showcase interesting objects from around the world, both historical and contemporary, but also have an excellent collection of costume that can be viewed upon request.

一些设计师喜欢通过服装来表达抽象概念（例如隔离），另外的设计师使用更形象的概念（例如圆圈）。

这两种方法都是可行的，就看哪种选择适合你。确实需要适合你的主题，选择一个不能使你兴奋的主题是毫无意义的。如果一段时间过后，理念仍然姗姗来迟，聪明的设计师就要坦诚地质疑所选择的主题是否合适。

记住，媒体和购买者只对结果感兴趣。那些衣服看上去漂亮吗？穿着它们使我更漂亮吗？它们令人激动吗？它们会有销售吗？他们没有必要关心你如何设法在一件夹克上表达量子物理学，但是如果这是你想表达的，那你就去做吧。

6.4 研究和资源

从哪里开始研究，这要根据你的主题或概念而定。对于好学的设计师来说，研究就像是侦探工作，去挖掘将会燃起火花的独特信息和主题材料。

开始研究最便利的地方是互联网。网络是图片和信息的神奇资源库。从网络也能直接了解面料制造商的信息，它们可能生产社会化大众面料，也有一些公司能够提供特别的服务。

一个好的图书馆也值得珍惜。图书馆提供的书籍往往考虑到社会不同行业的需求，因此，只有少量的专业书籍。专业图书馆最有收获，越是老的图书馆越好，往往有一些绝版的书籍仍然陈列在书架上，或至少根据要求能看到。学院和大学应该有一个根据所教专业配套相应图书的图书馆。但是如果你不在那里学习，进入图书馆将受到限制。

跳蚤市场和古玩市场也是十分有用的资源库，设计师可以从那里找到启发灵感的物品和资料。毋庸置疑，那里有各种各样的服装，古代的或当代的，能够带来更多灵感设计。例如，历史的、民族的、部队的制服，使你看到以前从未见过的细节、制造方法和结构。

和跳蚤市场一样，慈善商店也是寻找服装、书籍、光盘和小古董的好地方。在那里有时立即能找到灵感，有时只要稍加一点想象就能找到灵感。每天都有一些不再受欢迎或者被认为是劣质的东西被人们重新欣赏、再次发现和嘲讽地使用，运用到服装设计中。

博物馆，例如伦敦维多利亚和艾尔伯特博物馆，不仅收藏和展示来自世界各地的历史和当代的奇珍异品，而且还收藏了精湛的古代服装，提出申请就可以看到它们。

专业词汇

8. abstract adj. 抽象

通用词汇

⑱ quantum n. 量子
⑲ ignite vt. 点燃
⑳ antique adj. 古老的
㉑ ethnic adj. 种族的
㉒ kitsch n. 粗劣的作品

Large companies, with the budget, send their designers on research trips, often abroad, to search for inspiration. There, the designers are armed with a research budget and a camera, and can record and buy anything that might prove useful for the coming season. Designers with a tight budget might use a holiday abroad as a similar opportunity.

Sources of images can be photocopies, postcards, photographs, tear sheets from magazines and drawing. But anything can be used for research: images, fabric, details such as buttons or an antique collar[9]—anything that inspires you qualifies as research. Whichever items you collect must be within easy reach (and view) so that you have your reference constantly about you.

6.5 The research book

As a designer you will eventually develop an individual approach to "processing" this research. Some designers collect piles of photocopies and fabrics that may find their way on to a wall in the studio. Others compile research or sketchbooks where images, fabrics and trimmings are collected and collated, recording the origin and evolution of a collection. Still others take the essence of the research and produce what are called mood, theme or storyboards.

A research book is not necessarily solely for the designer's use. Showing research to other people is useful when trying to convey the themes of a collection. It might be used to communicate your concept to your tutor, your employers, employees or a stylist.

Research books are not just scrapbooks. A scrapbook infers that the information is collected, but unprocessed. There is nothing duller than looking through pages of lifeless, rectangular[10] images that have been (too) carefully cut out. It is also debatable[㉓] how much the designer has gleaned from creating pages like this. A research book should reflect the thought processes and personal approach to the project. It becomes more personal when it is drawn on and written in, and when the images and materials that have been collected are manipulated[㉔] or collaged[11].

6.6 Collage

The word "collage" is derived from French word for glue. A good collage is where the separate elements (images) work on different levels at the same time, to form both a whole and also its individual component parts.

Successful collage usually includes a bricolage[12] of different-sized, differently sourced images that provide a stimulating visual rhythm.

6.7 Drawing

Drawing a part or the whole of a picture you have collected as research helps you to understand the shapes and forms that make up the image, which, in turn, enables you to appreciate and utilize the same curve in a design or when cutting a pattern.

一些有预算的大公司提供设计师研究性的旅行，到国外去寻找灵感。到国外，设计师带着研究目的和相机，记录或者购买对下一季有帮助的东西。预算紧张的设计师也可以利用去国外度假，作为寻找灵感的机会。

图片资源可以是照片复印件、明信片、照片、从杂志和绘画上撕下的小碎片。但是可以从任何东西中进行研究：图片、面料、纽扣或者古代的领子等细节，任何能够激起灵感的东西都可以用来研究。不管你收集什么，必须是在能够得到的范围内，这样在你周围的参考资料源源不断。

6.5 研究本

作为一个设计师，最终你要形成一种个性化研究过程的方法。一些设计师也许找到了他们的方法，就是将收集的一大堆照片复印件和面料贴在工作室的墙上；还有一些设计师编辑所做的研究或草稿本，即将图片、面料和装饰品放在一起比对，记录一个系列的起源和发展过程。还有一些设计师汲取研究的精华，根据情绪、主题或故事板进行编辑。

研究不仅仅对设计师有用，当要告知别人系列的主题，用研究本传达信息显得尤其有效。你可以借助研究本将你的概念告诉你的导师、你的雇员、你的雇主或造型师。

研究本不是剪贴簿。一本剪贴簿意味着是信息的收集，没有发展进程。看着那些没有生气的、仔细剪下的长方形图片是很乏味的事情。有多少设计师这样做是值得商榷的。研究本应该反映从事这个项目的思想过程和个性化方法。如果在研究本上有你的绘画、你写的文字，并将收集的图片和资料拼贴在一起时，研究本就更加个性化。

6.6 拼贴

“拼贴”来源于法语“黏贴”一词。好的拼贴是同时将不同层面的各个元素和图形放在一起，形成一个整体，同时个体又各具个性。

成功地拼贴包括不同尺寸、不同来源的图片拼贴在一起，产生令人兴奋的视觉节奏。

6.7 绘画

在研究时，从你收集的某幅图片中画它的局部或全部，将帮助你理解形状和构图形式；反过来，在你设计或裁剪图形时，使你能够欣赏和利用相同的曲线。

专业词汇

9. collar n. 衣领
10. rectangular adj. 长方形的
11. collage vt. 拼贴
12. bricolage n. 拼装

通用词汇

㉓ debatable adj. 可争辩的
㉔ manipulate vt. 熟练控制

Using collage and making your own drawings allows you to deconstruct㉕ an image such as a photograph, photocopy, drawing or postcard. This is necessary because it may not be the whole image that will ultimately be useful to your designs; a picture may have been chosen for its "whole", but it is only when it has been examined in more depth[13] that other useful elements may be discovered. For example, a photograph of a Gothic cathedral is rich in decorative[14] flourishes[15], but it almost needs a magnifying glass to be able to understand the detail. By cutting up an image[16] or using a "viewfinder㉖" —a rectangle paper "frame" that enables you to focus on part of an image, much like the viewfinder on a camera—a rectangle paper smaller elements or details can become more apparent and be more easily examined.

6.8 Juxtaposition㉗

Placing images and fabrics together on the pages of your research book will help you to make important decisions about the content of your designs. Sometimes disparate images or materials may share similarities even though they are essentially different. For example, the spiral[17] shape of an ammonite fossil㉘ is similar to a spiral staircase or a rosette[18]. Or an image may be suggestive of a fabric you have sourced-for example, a place of velvet[19] evoke the texture of moss and lead you to think about natural imagery.

By utilizing drawing, collage and juxtaposition in your research books, you are processing and analyzing what has been collected. You are able to render and interpret images and materials as part of your own logical progression or journey.

6.9 Mood, theme and storyboards

Mood, theme and storyboards are essentially a distillation㉙ of research. In a sense they are the "presentation" version of the research book. They are made up as collage, and as the name suggests, generally mounted on board, which makes them more durable. They are used by a designer to communicate the themes, concepts, colors and fabrics that will be used to design the season's collection. They may include key words that convey a "feeling" such as "comfort[20]" or "seduction[21]". If the collection must be tailored to a particular client, the images may be more specifically attuned to the perceived lifestyle/identity of the potential client.

使用拼贴和自己的绘图，使你能够解构照片、复印图片、绘画和明信片这些图形。这样做是必要的，因为不是整幅图片对你的设计有用。一幅图片也许整幅都被选来，但是，仔细研究之后，可能会发现其他元素更有用。例如，一幅反映了辉煌装饰的哥特式教堂图片，几乎需要用放大镜才能理解它的细节。但是剪开一幅图片，或者使用取景器—— 一种长方形的框子，很像照相机的取景器，使你能够集中图片的局部，很小的元素或细节变得更加明显，更加容易审视。

6.8 并置

将图片和面料放在你研究本不同的页面上，将帮助你做出关于设计内容的重要决定。有时，即使它们在本质上完全不同的图片或资料也许有共同特点。例如，一个菊石化石的螺旋形状与楼梯或玫瑰花相似。或者一幅图片与你寻找的一块面料相似。例如，一块天鹅绒面料想起苔藓的肌理，使你想起自然的景象。

在你的研究本上使用绘画、拼贴和并置，处理和分析已经收集的资料，你将能够提供图片和资料的解释，使它们成为你逻辑进程的一部分。

6.9 情趣、主题和故事板

情趣、主题和故事板从本质上是研究的升华。实际上，是研究本的呈现版本。顾名思义，由于由拼贴构成，通常放置在一张页面上，使它们更具有持久性。设计师用它交流主题、概念、色彩和面料，这些将是设计师下一季设计时使用的。它们可能包括关键词，传达一种情感，例如舒适或者诱惑。如果这个系列是为一个特定的客户制作，图形将更加特别，要与潜在客户感知的生活方式和身份相协调。

专业词汇

13. depth n. 深度
14. decorative adj. 装饰
15. flourish n. 花样
16. image n. 影像
17. spiral n. 螺旋线
18. rosette n. 玫瑰花结
19. velvet n. 丝绒
20. comfort n. 舒适
21. seduction n. 诱惑

通用词汇

㉕ deconstruct vt. 解构
㉖ viewfinder n. 取景器
㉗ juxtaposition n. 并置
㉘ fossil n. 化石
㉙ distillation n. 蒸馏

6. 10 Designing

Once your research has been collated[30], you can start on design. But there is nothing more intimidating than a blank page. The process can be very frustrating; even when the designs start to come it can take a while before any of them are very satisfactory. This is a natural part of the design process. Many early designs are thrown away and you might even begin to question your abilities. Don't panic! It takes time to hit your stride, and after sweating a while over the page better ideas will start to emerge. Explore every possibility that comes to mind and discard nothing at this stage. You might discover the potential of an ideal later on when you look back over you designs.

A designer's identity or style comes with time, but as well as that, the clothed themselves need an identity or to form part of a vision in order to stand apart from the competition. Certain elements should run through the designs to give them coherence. It could be where an armhole[22] is cut, the placement of a seam[23] on the body in a particular way, or a method of finishing[24] the fabric. If these elements tie in strongly with your theme to work as a "whole" you are on your way to making a real statement with your designs.

6.10 设计

一旦研究完成，你可以开始设计。但是，没有什么比一张白纸更恐惧。这个过程可能是相当令人沮丧的，甚至设计开始以后的一段时间内，没有一个设计是满意的，这是设计过程的一部分。开始时扔掉很多设计，你甚至开始怀疑你的能力。不要惊慌！迈开步子需要时间，面对纸张出一会儿汗以后，更好的理念才会出现。在这个阶段探索每一种可能，不要放弃任何东西。当你回头看你的设计时，你可能发现某种具有潜力的理念。

设计师的特征或风格形成需假以时日，为了在竞争中脱颖而出，设计师需要有特征，或者将特征作为视觉构成的一部分。某些元素应该贯穿设计，使它们取得和谐。可能在袖窿处裁剪，用某种方式在身体某个部位放置一条缝线，或者面料的一种后处理方法。如果这些元素与你的主题紧密相连、构成整体，你的设计就真正体现出你的独到之处。

专业词汇

22. armhole n. 袖窿
23. seam n. 接缝
24. finishing adj. 修整

通用词汇

㉚ collate vt. 校对

Exercises

(1) Understanding the text.

Read the text and answer the following questions.

1) What are the ethical standards observed by the fashion design community?

2) Can you analyze one famous clothing brand?

3) Please choose products for fashion design.

(2) Building your language.

Translate the following sentences into Chinese.

1) Novelty belts can be made of any material needed to meet fashion requirements: webbing, plastic, braid, chain, rope, or even rubber.

____

2) Shoe companies may have shoe lines in all price ranges and categories.

____

3) American designer score high when it comes to marketing savvy and making salable clothes that appeal to the whole U. S. population.

____

4) Canada and Mexico are the largest trading partners of the United States, and in turn, the United States account for more than two-third of their total trade.

____

5) Social apparel includes special occasion attire such as long and short cocktail dresses, dressy pant ensembles and evening and bridal gowns.

____

(3) Sharing your ideas.

Please write a short introduction (of around 300 words) about fashion design. Try to make full use of what you've learned from text, including the relevant information from the reading text as well as words and expressions.

Unit 7

Fashion Designer

第7单元

时尚设计师

Are you thinking of becoming a fashion
designer? Like Halston, Lagerfeld you'll be
creating way more than just art - you'll be
creating art people live their lives in.
And while some may think fashion and
appearance are superficial, you know that
what you do helps people present themselves to
the world, get hired, impress a first date and feel
great about themselves. In reality, there a[illegible]
ton of different positions within fashion design
itself. From workout wear to haute couture,
someone somewhere had vision and designed it.
But it's [illegible] sc[illegible]ing at runway?
Shows in [illegible]k, it's also
a lot of hard [illegible]st definitely
have to pay your d[illegible] you show under
the big tents3. Fash[illegible]ners must excel in
creativity4, style, kn[illegible] of fashion history
business sense, and th[illegible] to perform under
pressure and in high[illegible]vironments.
So how do you become a fashion designer?
1. Honing your fash[illegible]n design skills
Successful fashion designers have a wide

Are you thinking of becoming a fashion designer? Like Halston and Lagerfeld, you'll be creating way more than just art— you' ll be creating art people live their lives in.

And while some may think fashion and appearance[1] are superficial, you know that what you do helps people present themselves to the world, get hired, impress a first date and feel great about themselves. In reality, there are a ton of different positions within fashion design itself. From workout wear to haute couture, someone somewhere had vision and designed it.

But it's not all schmoozing at runway[2] shows in Paris and New York—it's also a lot of hard work and you'll most definitely have to pay your dues before you show under the big tents[3]. Fashion designers must excel in creativity[4], style, knowledge of fashion history, business sense, and the ability to perform under pressure and in high stress environments. So how do you become a fashion designer?

7.1 Honing your fashion design skills[5]

7.1.1 Develop your skills

Successful fashion designers have a wide array of skills, including drawing[6], an eye for color and texture, an ability to visualize① concepts[7] in three dimensions[8], and the mechanical skills involved in sewing[9] and cutting[10] all types of fabrics.

(1) Get excellent sewing tuition if you haven't already learned this skill well. Being able to sew difficult fabric under challenging situations will stand you in excellent stead throughout your career but you need to work at it—it's a skill that doesn't come easily to many people.

(2) Understand how fabrics move, drape[11], breathe, react when worn, etc. Your in-depth knowledge of fabric is absolutely essential to using it properly when designing. Also know where to source materials[12] from.

(3) Learn from existing designers, not just who they are, but their backgrounds, their signature style, the learning that they undertook, where they studied. Knowing this will help you to be a better designer yourself, as you can borrow and build on their ideas.

(4) Learn how to create storyboards② and product ranges[13]. Be good at researching trends through media, comparative shopping and trade shows.

(5) Start developing these skills at a young age. Be prepared to devote hours of time to perfecting your craft.

7.1.2 Learn more

If you can, it makes good sense to get a diploma or degree in fashion design or a related program. You'll learn a great deal, make excellent early contacts and have ample opportunity to show off your skills in a less judgmental environment (although still be prepared to be critiqued)! Do one (or both) of the following:

你觉得你能成为一名时尚设计师吗？要想成为候司顿（Halston）、拉格菲尔德（Lagerfeld）那样的设计师，你就要采用创造性的艺术方式——你要创造人们生活中的艺术。

虽然人们认为时尚和外貌是表面的，你知道你是在帮助人们向世界介绍他们自身，帮助他们找到工作、留下好的第一印象和充满自信。在现实中，时尚设计领域有大量的工种。从运动服到高级时装，要根据不同的人、不同的地点设计不同的服装样式。

但是，这不是在闲聊巴黎和纽约的时装发布会——那是太多的辛苦工作，在你的作品能够在大帐篷里展示之前，毫无疑问你必须付出艰辛的努力。时装设计师必须在创意、风格、时尚历史知识和商业意识方面精益求精，要有承受工作压力和竞争环境的能力。因此，如何成为一名时尚设计师呢？

7.1 磨炼你的时尚设计技能

7.1.1 发展你的技能

成功的时尚设计师要有多方面的技能，包括绘画、对色彩和肌理的审视力，在三维空间中将概念视觉化的能力，缝纫和裁剪各种面料的操作能力。

（1）如果你还未学会缝纫，到一个优秀缝纫班学习。在挑战性的环境中如果你能熟练地缝制有缝纫难度的面料，在你的整个职业生涯中你就能立于不败之地，因为很多人不能轻易掌握这种技能。

（2）理解面料穿在人体上的动感、悬垂性、透气性和反应等。你对面料有很深的了解，这很重要的，因为你将能够在设计中知道如何正确地使用。还要知道面料来自何处。

（3）向现有的设计师学习，不仅知道他们是谁，还要知道他们的背景，他们独特的风格，他们采取的学习方法如何，在哪里学习。知道了这些将帮助你成为一个更好的设计师，因为你可以借用并以他们的理念为基础。

（4）学习如何创建故事板和产品类别。善于根据媒体、销售对比和展销会研究流行趋势。

（5）在年轻的时候开始发展这些技能，并花一些时间完善你的技能。

7.1.2 更多了解

如果可以的话，最好获得时装设计或相关课程的文凭或学位。你会学到很多东西，有优良的开端，在一个不太根据技能判断能力的环境中显露下你自己（虽然仍要准备接受批评）！按照下面其中一项（或两项）操作：

专业词汇

1. appearance n. 外貌
2. runway n. 跑道
3. tent n. 帐篷
4. creativity n. 创造性
5. skill n. 技能
6. draw vt. 绘画
7. concept n. 概念
8. dimension n. 维度
9. sew vt. 缝纫
10. cut vt. 裁剪
11. drape vt. 悬挂
12. material n. 材料
13. range n. 类别

通用词汇

① visualize vt. 使可见

② storyboard n. 故事板

(1) Get a degree in fashion design. Most programs are three or four years long. FIDM and Parsons are two of the most popular design schools in the United States. You will study drawing, color and composition[14] pattern[15]-making, and draping. In addition to learning practical skills like these, you will also be working with industry professionals who may serve as important contacts in the future and who can give you first-hand advice and feedback on your work.

(2) Apply for an internship③ or apprenticeship④. If school is not for you, or if you simply feel that real-world experience will be of more benefit to you, then find a fashion internship. You will need to have an impressive portfolio⑤ to apply and be willing to start at the bottom; interns are often given menial tasks like getting coffee. Again, the connections you make through your internship or apprenticeship will be vital as you pursue your career in fashion, and working with industry professionals will give you an opportunity to pick up important skills first-hand.

7.2 Working out which fashion is your passion

7.2.1 Decide which designing field is your principal interest

You may need to start at the bottom but you do need to have a goal in mind as to the type of designing you want to undertake lifelong. Are you interested most in haute couture, ready-to-wear[16], fitness[17], leisure[18] gear, the mass market or niches such as eco wear? Each has advantages and disadvantages that you'll need to explore before reaching your final decision on which pathway to pursue. Within these major fields, you'll also need to decide on a few sub-set areas for your fashion design. You might wish to straddle a few but to begin with, don't over-extend yourself as it's better to perfect your designing within one area and then experiment when you've already got a good foothold in the industry.

For Example:

- Women's daywear[19], women's evening wear.
- Men's daywear, men's evening wear.
- Boys' wear and/or girls wear; teenage wear.
- Sportswear[20]/fitness/leisure wear.
- Knitwear[21].
- Outdoor, adventure, outerwear[22].

（1）获得一个时尚设计文凭。大多数课程有三年或四年时间。FIDM 和帕森斯（Parsons）是美国最著名的两个设计学校。你将学习绘画、色彩和结构、样板以及立体裁剪。除了学习这些实践技能以外，你还将与行业的专业人士一起工作，他们可能成为你未来的重要联系人，他们可能给你的工作提出第一手意见和反馈。

（2）申请实习生或学徒。如果学校没有为你安排，或者你觉得实际经验对你有更多的好处，那么你可以去做一个时尚实习生。在申请时，你需要有引人注目的申请材料，而且愿意从底层开始做起；实习生经常被叫去做一些打杂的活，例如泡咖啡。同样，在实习或学徒建立的关系对你寻求时尚领域的职业很有用，与行业专业人士一起工作将使你获得重要的第一手技能。

7.2 弄清你热爱哪种时尚

7.2.1 决定哪个设计领域是你的主要兴趣

你可能要从底层做起，但是在你的内心你需要有一个目标，哪种类型的设计是你想终身从事的。你对高级时尚感兴趣，还是成衣、健康服、休闲服装、大众或小众市场，例如生态服装，每一种服装有优势和劣势，在你做出你追求目标的最终决定之前，你要好好地研究。在这些大的分类里，你还要决定你将从事哪个子类的时尚设计。你也许希望脚踩几条船，但是开始的时候，不要跨得太多，因为在一个领域内完善你的设计是比较好的，当你在这个行业立足之后你再尝试其他领域。

例如：

· 女性日装，女性晚装。
· 男性日装，男性晚装。
· 男孩服装，或女孩服装；青年人服装。
· 体育运动服装/健康服装/休闲服装
· 针织服装。
· 户外服装，探险服装，外衣。

专业词汇

14. composition n. 组合方式
15. pattern n. 纸样
16. ready-to-wear adj. 成衣
17. fitness n. 健康服
18. leisure n. 休闲服
19. daywear n. 日装
20. sportswear n. 运动装
21. knitwear n. 针织品
22. outerwear n. 外衣

通用词汇

③ internship n. 实习
④ apprenticeship n. 学徒
⑤ portfolio n. 文件夹

· Bridal[23] wear.

· Accessories.

· Costume design for theater, movies, the advertising industry and retailers.

7. 2. 2 Plan some key pieces

What is your absolute strength in designing? Perhaps you're a whizz[⑥] at accessories or a genus with yoga pants[24]. Your passion and skill are an important first part of the equation[25]. Of course, the second part is matching this to what the market wants, which in fashion, is part convincing the market and part noticing what the market is demanding.

7. 3 Deciding if the fashion industry is ready for you

Assess your skills and personality honestly before pursuing a career in fashion design. You may love clothes but clothing is only part of the story when undertaking fashion design. You'll also need excellent communication skills, a willingness to work very hard, a tough hide when criticized, an ability to cope with stress, openness to having many different clients[26] and/or bosses, an acceptance that there will be loneliness or isolation[⑦] on occasion (depending on how you set up your design business or career) and an ability to be a self-disciplined self-starter.

Fashion product is the fashion-buying public to decide, and for every person who likes your work there will be someone who really doesn't. This is common and working in such a subjective field can be confusing, you need to learn to navigate[⑧] your way through criticism[⑨] and either develop a steely exterior of recognize which opinions you respect and which to disregard. Once you accept this, you are free to get on with you are best at designing clothes.

Being a fashion designer is probably for you if: You want to devote your life to this career (it's your "vocation"), you don't mind uncertainty or insecurity, you are willing to stand up for what you believe in, you have distinct ideas about what is important in fashion, you listen to clients well, you know the fashion industry inside out and you live, eat and breathe fashion.

Being a fashion designer is probably not for you if: You can't manage stress well you don't like uncertainty or instability, you want a career without too many highs or lows, you need other people to praise your efforts, you need a lot of guidance, you hate being financially unstable and you have too many other interests in life.

· 新娘服装。

· 配饰。

· 戏剧服装，电影服装，广告服装和销售人员服装。

7.2.2 计划关键实力

设计中什么是你的绝对实力，也许你在配饰上是高手，或设计瑜伽裤子是天才。你的激情和技能是人生等式的第一个重要组成部分。第二部分是将激情和技能与市场需求相吻合。在时尚中，一部分是说服市场，另一部分是注视市场正需要什么。

7.3 决定你是否已经准备好进入时尚行业

在追求时尚设计职业前诚实地评估你的技能和个性。你可能喜爱服装，但是服装只是从事时尚设计的部分内容。你还需要优良的沟通技能，乐意刻苦地工作，受到批评时有坚强的忍受力，应对压力的能力，友善地对待不同客户和（或）老板。忍受某些时候的孤独或隔离（取决于你如何设置你的设计业务或职业），有自我约束和积极主动的能力。

时尚产品由公众的时尚购买决定，有人喜欢你的作品，也有人不喜欢，这是正常的。工作在这种主观性行业里容易搞混，你需要根据批评意见学会驾驭自己，建立铁石心肠般的外表，识别哪种是值得你尊重的意见，哪种是需要舍弃的意见。一旦你习惯这样做，你可以自由地从事服装设计。

你要想成为一名时尚设计师，如果：你想将你的一生投入这个职业（这是你的“天职”），你不在乎不稳定或不安全，你愿意坚持你的信仰，你对什么是时尚中的重要部分有独特的理念，你虚心听取客户的意见，你知道时尚行业的里里外外，你与时尚共生命。

你将不是一名时尚设计师，如果你不能承受压力，你不喜欢不确定或不稳定，你不想一种职业有太多的起伏，你需要他人表扬你的努力，你需要很多的指导，你很不喜欢经济上不稳定，你的生活中有其他太多的兴趣。

专业词汇

23. bridal adj. 新娘的
24. pants n. 裤子
25. equation n. 人生等式
26. client n. 顾客

通用词汇

⑥ whizz n. 专家
⑦ isolation n. 隔离
⑧ navigate vt. 驾驭
⑨ criticism n. 批评

7.4 Setting yourself up for success

7.4.1 Get educated about the business side of fashion

Being a successful fashion designer not only requires talent and creativity, it also requires a sound knowledge of the business and marketing aspects of the fashion world. Keep yourself updated on the happenings in the fashion industry by regularly reading trade journals like Women's Wear Daily and Daily News Record.

(1) Many fashion design programs include courses in marketing. Some programs/majors highlight marketing more than others, so be sure to do ample research on the coursework involved in the program you choose. If you've already undertaken a course but missed the marketing/financial side of things, consider doing short courses in these aspects of business.

(2) Learn beyond design. There is an entire supply chain[27] involved in the fashion industry and you need to understand what each persons job is, so that you can see things from their perspective too, in order to make compromises, meet demands and understand where things get held up. Research what others do, such as buyers, merchandisers[⑩], pattern cutters, garment and fabric technologists, quality controllers, graders[28], sample machinists, sales people, PR and marketing people, fashion journalists, retailers, event organizers, fashion stylists[29] and so forth.

(3) Know your customer. This skill is basic and essential[⑪] and it's one a fashion designer must never lose sight of. Know how much your customers spend, what their lifestyles are, where they like to shop, how they like to shop and what they like and dislike. Knowing what are absolute needs and what are the things that only get bought when disposable[⑫] incomes are less tight. If you have done marketing, you should have a solid understanding of how to work out customers' needs.

(4) Know your competitors. Always keep an eye on what other fashion designers in your area of interest are doing. At a minimum[30], keep up. Better still, surpass[⑬] them while still meeting your customers' needs.

(5) Trade fairs are an excellent place to develop deeper understanding of how the fashion industry works and what will work for you in terms of meeting customer needs and staying competitive.

7.4.2 Look for fashion design jobs

There are various ways to find work in the fashion industry as a designer and it depends on the type of designing you're interested in. In some cases, being versatile[⑭] will help you a great deal, just so that you get the experience and then jump across to your real passion later. And in most cases, you'll need to be persistent[⑮] and apply to many different places to get your foot in the door. For starters, some places to apply to include:

7.4 迈向成功

7.4.1 接受时尚经营方面的教育

要想成为一名成功的时尚设计师，不仅需要天才和创造性，还需要具备时尚经营和市场方面的丰富知识。经常地阅读时尚行业方面的期刊，例如 Women's Wear Daily 和 Daily News Record，以便及时了解时尚行业内发生的事情。

(1) 许多时尚设计科目包括营销课程，一些课程或专业更注重营销，因此要充分地研究你选择的科目所涉及的课程。如果你已经进行了课程的学习，但错过了营销课程或金融方面知识的学习，要考虑学习一些短期课程。

(2) 学习设计以外的知识。在时尚行业有一条完整的供应链，你需要了解每一个人做的工作是什么，以便你能够从他们的角度看问题，才能委曲求全、满足需求和知道在哪里受阻。研究其他人做什么，例如买手、跟单员、样板裁剪师、服装和面料技术专家、质量控制员、放码师、样衣师、销售员、公共关系（PR）和营销人员、时尚新闻工作者、零售商、活动组织者和时尚造型师，等等。

(3) 了解你的客户。这是基本和重要的能力，也是时尚设计师不可忽视的。了解你的客户花多少钱，他们的生活方式，他们喜欢在哪里消费，他们喜欢怎样购物，他们喜欢什么，不喜欢什么。知道什么是绝对的需求，什么是可支配收入，不太紧张时才会购买的东西。如果你已经做营销，你必须深刻地了解如何满足客户的需求。

(4) 了解你的竞争者。不断观察在你感兴趣的时尚领域内其他时尚设计师在做什么，至少跟上他们。更有甚者，超越他们的同时还能满足客户的需求。

(5) 展销会是帮助深入了解时尚行业运作的最好地方，你将会知道你做什么能够满足消费者的需求和保持竞争力。

7.4.2 寻找时尚设计工作

在时尚行业，有很多方法可以找到设计师的工作，这取决于你对哪种类型的设计感兴趣。在一些情形下，做各种类型的工作将会得到很多经验，然后再跳到你真正有激情的地方。在大多数情形下，你需要耐心地申请很多不同的地方，才能找到工作。对于一个刚工作的人来说，可以应聘的地方包括：

专业词汇

27. supply chain 供应链
28. grader n. 放码师
29. stylist n. 造型师
30. minimum n. 最小量

通用词汇

⑩ merchandiser n. 商人
⑪ essential adj. 本质的
⑫ disposable adj. 可任意处理的
⑬ surpass vt. 超过
⑭ versatile adj. 多用途的
⑮ persistent adj. 坚持不懈的

(1) Existing fashion houses and designers—look for internships, entry-level paid positions, assistants to designers, etc.

(2) Costume positions with movie studios, theaters, costume stores, etc.

(3) Online advertisements through various online job agencies.

(4) Word of mouth—use your college or fashion industry contacts to get you through the door. In an industry that values what people who already are well positioned have to this is a good way to get started.

7.4.3 If running your own design business, be prepared to be financially astute

You may be exceptionally⑯ creative but be absolutely certain that if you run your own fashion label[31], you need to be business savvy⑰. You do need to understand those numbers and the invoices⑱ that keep piling up on your table. If you really hate this stuff, there are good options, such as asking your accountant to take care of all things financial but it still pays to keep on top of the whole thing yourself. And if you really, really hate this side of it, look for work as a fashion designer with a fashion house instead of running your own label.

What type of trader will you be? There are many possibilities, including sole trader, partnership, incorporated⑲ company, etc. Each has distinct advantages and disadvantage that you should discuss with your legal and financial advisers before proceeding. Be sure that you are covered for liability in all circumstances, especially if you're in a particularly litigious⑳ culture.

7.4.4 Be realistic

You may need to be willing to move to match your market but that depends on how you work and sell. Being realistic means recognizing that it's pointless to trying sell a lot of haute couture to people who only wants career clothing in a semi-rural town while it's no good trying to sell bikini[32] to the Inuit. You'll need to focus on where your market is most likely to be and either work out whether it's best for you to live and work in that same area or how to get the distribution㉑ from your current area to the place where it's most likely to sell.

(1) Take into consideration the influences around you. As a creative person, part of your creative process is being around like people and sparking off their ideas and suggestions too. It's a lot harder to do this alone or working alongside people who aren't into your fashion approaches.

(2) Remember too that seasonality impacts fashion designing and may have an impact on the type of clothing you're producing and where you wish to sell it.

（1）现有的时尚屋和设计师品牌公司——寻找实习生、初级薪水的职位、助理设计师，等等。

（2）电影工作室、剧院和戏剧服装商店设计戏剧服装等。

（3）通过各种工作中介网站发求职广告。

（3）口头传播——通过与你所在学院或时尚行业取得联系，帮助你找工作。在行业中，如果有一个好职位，不得不说是好的开始。

7.4.3 如果你想从事自己的设计业务，要在金融上很精明

你可能非常有创意，但如果你运作自己的时尚品牌，你绝对需要商业头脑。你需要明白那些不断堆放在你桌子上账务和发票。如果你不喜欢这些，有一个很好的选择，请你的会计处理金融方面的所有事务，但是，你仍然要总管所有的事务。如果你非常不喜欢这些事务，就在时尚屋里寻找一份工作，而不要经营自己的品牌。

你将是哪种类型的商人？有很多种可能，包括专营商、合作商、股份公司，等等。每一种都有自身的优势和劣势，在工作进展之前，你必须和你的法律或金融顾问讨论，确保在任何情况下你受到法律的保护，特别是如果你在一个容易产生诉讼的情形下。

7.4.4 求实

根据你的工作状况和销售，你可能需要主动地调整去适应你的市场。求实意味着如果将高级时装销售给需要职业服装的半乡村城镇人，是毫无意义的；同样，将比基尼销售给因纽特人（Inuit）也是无意义的。你需要把重点放在最有可能的市场，既要了解你是否适合生活和工作在同一个区域，又要考虑从你工作的地点到销售最好地点的配送问题。

（1）考虑你周围的影响。作为一名设计者，你设计过程中的一部分就是和你喜欢的人一起，从他们的理念和建议中燃起火花。单独工作或不能和你的工作方法合拍的人一起工作是很困难的。

（2）记得时尚设计受季节性的影响，也许会影响你正在生产的服装品种，或者影响你希望销售的区域。

专业词汇

31. label n. 标签　　32. bikini n. 比基尼

通用词汇

⑯ exceptionally adv. 例外地
⑰ savvy n. 悟性
⑱ invoice n. 发票
⑲ incorporated adj. 股份有限的
⑳ litigious adj. 好争论的
㉑ distribution n. 分配

(3) Consider the power of online selling. Provided you use good quality three dimensional images that can be zoomed and turned, selling your fashion online to anywhere in the world is another realistic possibility nowadays. This allows you greater flexibility[33] in where you'll live and design and can reduce the daily commute㉒ to zero. This may be ideal if you plan on staying a small fashion label. Even then however, you should still make allowances[34] for traveling to major fashion shows.

(4) Living in a city with a thriving fashion industry makes good sense for many designers. According to the Global Language Monitor (GLM), the following cities were the top fashion capitals of the world in 2012, in descending order:

1) London, England.

2) New York, US.

3) Barcelona, Spain.

4) Paris, France.

5) Mexico City.

6) Madrid, Spain.

7) Rome, Italy.

8) Sao Palo, Brazil.

9) Milan, Italy.

10) Los Angeles, US.

11) Berlin, Germany.

7.5 Creating your fashion portfolio

Assemble[35] a portfolio of your work. Your design portfolio will be vital when applying to design jobs and internships, as it is your chance to market yourself and your work. Your portfolio should display your best work, and highlight your skills and creativity. Use a high quality binder to show that you take yourself seriously as a designer. Include the following in your portfolio:

(1) Hand-drawn sketches[36] or photographs of these sketches.

(2) Computer-drawn designs.

(3) Resume.

(4) Mood or concept pages.

(5) Color or textile presentation pages

(6) Any other pieces that fairly reflect what you're capable of doing and evolving into.

（3）考虑网络销售的威力。你要提供清晰的可以缩小和放大的三维图像，在网络上将你的产品销售到世界各地，这是当今另一种真实可能的销售方式，给你生活和设计带来更大的便利性，可以将日常往返费用减少到零。如果你计划小型的时尚品牌，这是理想的方法。即使那样，你还要有足够的费用去看大的时尚发布会。

（4）居住在时尚行业繁荣的城市，对于设计师来说能够培养好的感觉。根据全球语言检测机构［the Global Language Monitor（GLM）］统计，2012 年以下城市是世界顶级时尚首都，排名如下：

1）英国伦敦。

2）美国纽约。

3）西班牙巴塞罗拉。

4）法国巴黎。

5）墨西哥市。

6）西班牙马德里。

7）意大利罗马。

8）巴西圣保罗。

9）意大利米兰。

10）美国洛杉矶。

11）德国柏林。

7.5 设计你的时尚档案

装订你的作品档案。当你应聘设计工作和实习生时，你的设计档案是非常重要的，因为它是推销你自己和你作品的机会。你的档案必须展示你最好的作品，显示你的技能和创造力。使用高质量的黏合剂，说明你很在乎你成为一名设计师。包括如下的文件：

（1）手画的草图，或者草图的照片。

（2）计算机画的设计稿。

（3）简历。

（4）语境和概念页。

（5）色彩和面料展示页。

（6）任何其他能够很好地反映你的才能和进步的材料。

专业词汇

33. flexibility n. 伸缩性
34. allowance n. 补贴
35. assemble vt. 装配
36. sketch n. 草图

通用词汇

㉒ commute n. 通勤

Tips:

(1) If you're thinking of showing people your fashion drawings, think how you would look in your fashion drawings. Wear your own fashions as much as possible. What better way to promote your clothing than to wear it? When people ask questions about it, be ready to explain everything in short, pithy ways that excite the listener.

(2) Develop a good logo if running your own fashion label. It will define your style from the outset and so it needs to be good from the outset. It is worth getting a professional graphic[37] designer on the job if you're no good at this yourself.

(3) If you start your own fashion label, you need sound advice on everything from the beginning. Surround yourself by a trusted team of financial, legal and marketing advisers, paid according to what you need rather than having them on staff.

(4) It helps to be creative with your designs by adding color.

(5) Learn early on how to pack a decent lunch and snacks. Hours can be very long in fashion design and sometimes leaving your creativity zone may be impossible. Your brain needs good nutrition㉓ though, so by remembering to pack healthy lunches and snacks, you can grab something to sustain all that hard intellectual slog and physical running around without starving yourself silly.

(6) Read widely. Find the biographies and true stories of fashion icons in the area of fashion that you're interested in. Learn all of the ins and outs of their experiences and now you can use their experience to better your own. For example, if you want to shift into eco fashion, there are plenty of good trailblazer㉔ designers whose experiences have been documented.

(7) Be able to take insults. Nobody is perfect. Take advice from friends and family. Never give up, you can't quit your passion!

Warnings:

(1) The fashion industry is extremely competitive. Only pursue a career in fashion if it you are 100 percent[38] devoted to the field.

(2) Designing for catwalks and high end fashion will bring you into direct contact with the challenging aspects of the industry, including using underweight models[39] for fitting[40], cattiness from fellow designers and fashion industry elites and very difficult demands including tight deadline㉕. If you're not already an assertive person, it would be wise to spend time improving your skills in communicating and standing up for your principles.

(3) Working as a designer can be a physically strenuous㉖ career. You will need to be willing to work unexpected long hours to meet deadlines.

提示:

(1) 如果你想展示给别人看你的时尚绘画，首先你得考虑你对那些绘画的感觉如何。你自己尽可能穿着时尚。有什么方式比穿衣更好地推销你呢？当有人向你提问时，要用简练的话语解释，简练方式使听者高兴。

(2) 如果你运营自己的时尚品牌，设计一个好的 logo。它将从开始就很好地定义你的风格，一切需要从好的开始。如果你不擅长绘画，雇佣一个好的专业图形设计师是很值得的。

(3) 如果你开始从事自己的时尚品牌，需要听取多方面意见。你周围要有可信赖的团队，包括金融、法律和市场顾问，根据你的需要付给他们工资，而不是将他们作为员工。

(4) 添加色彩，将使你的设计更富有创意。

(5) 及早地学会如何包裹体面的午餐和点心。有时可能持续好几个小时进行时尚设计，将设计一半放在那里是不可能的。你的大脑需要良好的营养，因此记住要带些健康的午餐和点心，你可以吃些支撑辛苦智力活的食物，保证体力正常运行，而不至于傻傻地挨饿。

(6) 广泛地阅读。寻找你感兴趣的时尚偶像的传记和真实故事。了解他们的人生经历，考虑如何利用他们的经验改善自己。例如，如果你想转到生态时尚，有很多好的先锋设计师，他们的经验已经编撰成册。

(7) 能够承受侮辱。人无完人，采纳朋友和家人的意见。从不放弃你的激情。

警示:

(1) 时尚行业竞争十分激烈；追求时尚职业生涯需要你 100% 的努力。

(2) 设计走秀和高端时尚，将使你直接接触到行业内最具挑战性的方面，包括使用瘦型模特试衣，接触同行设计师和时尚界的精英，并且需要在很短时间内完成艰难的任务。如果你还缺乏主见，花些时间提高你的交流能力和坚持你的原则是明智的。

(3) 设计师的工作是耗费体力的职业，你需要主动地加班加点完成工期。

专业词汇

37. graphic adj. 图示的
38. percent n. 百分比
39. model n. 模特
40. fit vt. 试衣

通用词汇

㉓ nutrition n. 营养
㉔ trailblazer n. 先驱者
㉕ deadline n. 最后期限
㉖ strenuous adj. 费力的

Exercises

(1) Understanding the text.

Read the text and answer the following questions.

1) What's the image of fashion designer in your eyes?

2) Since fashion design is a highly competitive and ever changing industry, would you plesse to be a fashion designer?

3) How can you keep your horizons wide open when you consider something to design?

(2) Building your language.

The following expressions can be used to talk about fashion designer. Choose the right ones to fill in the blanks in the following sentences. Change the form where necessary.

passion	perseverance	exclusive	discerning
break into	rent	mental	designers

Have you always dreamed of becoming a fashion designer? Although it can be a difficult industry to ________, those that have the desire and the ________ can work their way into the field. Fashion design is a highly competitive and ever changing industry, and to succeed you will need drive and ________. Don't think you can't get into fashion design just because you aren't in New York or didn't graduate from an ________ fashion design school.

When you think of a fashion designer, you may have a ________ image of someone who works exclusively with expensive lines of clothing that are sold in only the most ________ stores. Although there are plenty of ________ that do this kind of work, you have to keep in mind that any kind of clothing must be designed. The clothes that you wear every day were created by a fashion designer. Just because something doesn't cost more than your ________ doesn't mean it didn't go through an extensive design process.

(3) Sharing your ideas.

After learning about fashion designer in this section, are you eager to share your knowledge? Please write a short introduction (of around 300 words) about fashion designer. Try to make full use of what you've learned from Text, including the relevant information from the reading text as well as words and expressions.

Drawing may be described as an evolutionary ① process that is fundamental to communicating ideas. This is also true of fashion drawing, with its distinctive nuances and associations with style. The exciting breadth1 and diversity ② of what constitutes fashion drawing today is testimony to the creative vision of fashion designers and fashion illustrators2 alike. It reflects the range and scope3 of media now available, from a simple graphite ③ pencil to sophisticated CAD programs.

Fashion [illegible] the act of drawing [illegible] also [illegible] of con[illegible] the page [illegible] have an idea [illegible] may sound [illegible] should be purpos[illegible]. In many [illegible] a fashion sketch is problem solving process which brings together the visual elements [illegible] an idea [illegible] purest form. This can mean recording a sudden moment

Unit 8

Fashion Drawing

第8单元

时尚绘画

Drawing may be described as an evolutionary① process that is fundamental to communicating ideas. This is also true of fashion drawing, with its distinctive nuances and associations with style. The exciting breadth[1] and diversity② of what constitutes fashion drawing today is testimony to the creative vision of fashion designers and fashion illustrators[2] alike. It reflects the range and scope[3] of media now available, from a simple graphite③ pencil to sophisticated CAD programs.

8.1 The sketching process

Fashion sketching not only involves the act of drawing an initial idea but also the process of considering and developing the idea across the pages of a sketchbook. It is always best to have an idea of what you want to draw. This may sound obvious, but fashion sketching should be purposeful, not random[4] or too abstract. In many respects a fashion sketch is a problem-solving process, which brings together the visual elements of articulating an idea in its purest form. This can mean recording a sudden idea before it is lost or forgotten, or capturing a moment in time, such as observing a detail on someone's garment.

A fashion sketch should seek to record and make sense of an idea. This is largely achieved with any one or more of three components: establishing the overall silhouette of a garment or outfit; conveying[5] the style lines of a garment such as a princess seam[6] or the positioning of a dart[7], and representing details on a garment such as a pocket[8] shape, top stitching[9] or embellishment. Some sketches may appear spontaneous④ or similar to mark making but they should all be linked by a common understanding of the human form and an end use.

Graphite or drawing pencils are ideal for shading and creating variations of line quality. While this is a good way to get started, it is also well worth developing the confidence to sketch with a pen. Sketching in pen requires a more linear[10] approach to drawing, which can often enhance the clarity[11] of a design idea, and it is no less spontaneous than using pencil.

8.2 Working drawings

In fashion it is quite usual to produce a series of rough sketches or working drawing in order to arrive at a design or collection proposal. This allows the designer to develop variations on an idea, before making a final decision about a design, whilst at the same time forming part of a critical process of elimination and refinement. The process of reviewing and refining a design involves collating ideas in line-up sheets[12]. These represent drawings of outfits (not individual[13] garments), which are visually presented on the human figure as a coherent statement for a collection proposal. Line-up sheets are more practical than inspiration sketches or rough sketches and are generally clearer to understand on the page. Their primary purpose is to assist with visual range planning and the commercial requirements of formulating[14] ready-to-wear clothing ranges. Consequently, they have no real basis in haute couture or bridal wear, which is more about representing the individual.

绘画可以被描述为渐进的过程，具有交流理念的功能。时装画以它独特的细微特征和风格实现其交流功能。今天，各种类型的时尚绘画令人激动，证明了时尚设计师和时尚插画家创造性的构想，也反映了应有尽有的绘画媒介，从简单的绘图铅笔到多功能的CAD软件。

8.1 草图过程

时尚草图不仅包括将最初理念画出来的行为，还包括在速写本上、不同页面画上思考和发展理念的过程。你总是有想画的想法非常好，这听起来显而易见，但是时尚草图应该有目的，不要随意或太抽象。从很多方面看，时尚草图就是解决问题的过程，以其最纯粹的形式，集中视觉元素，阐明一种想法，可以记录突发灵感，以免丢失或忘记，或者及时捕捉一瞬间，例如观察某人服装上的细节。

时尚草图就是寻求理念，记录下来，使理念富有含义，草图上包含一到三方面内容。它们是：确定一件服装或一套服装的整体轮廓；表达服装的款式线条，如公主线或省道的位置；表现服装的细节，例如口袋形状，表面线迹或装饰。一些草图可能做一些无拘束或相似的标记，但是这些关于体型的标记要使大家都能看懂，直到后期都能使用。

石墨或绘画铅笔是理想的工具，可以画阴影和各种形式的线条。在开始时就用钢笔是一种好方法，很适合建立绘画的自信。钢笔草图采用线描手法，使设计理念更加清晰，其即兴创作的效果与铅笔一样。

8.2 工作绘画

在时尚中，通常要画一系列潦草的草图或工作绘画，目的是得到一种设计和一个系列的提案，便于设计师做出设计最终决定之前开拓多种理念，与此同时，也是一种取消和修改设计的重要过程。设计的回顾和修改过程需要在图片旁边配上一张纸条，上面写着关于理念的文字。工作绘画要画整体着装状态（而不是单件服装），用人体形象表示，为系列提案配备文字说明。配上小纸条要比灵感草图或大概的草图来得实际，更容易理解页面上的内容。它的主要目的是帮助制定款式计划和根据商业需求制定成衣的款式计划。而说到高级时装或新娘服装，就不需要这些，因为它们要体现更多的个性特征。

专业词汇

1. breadth n. 宽度
2. illustrator n. 插图画家
3. scope n. 范围
4. random adj. 任意的
5. convey vt. 传递
6. princess seam 公主线
7. dart n. 省道
8. pocket n. 口袋
9. top stitching 面上切线
10. linear adj. 直线的
11. clarity n. 清楚
12. sheet n. 纸
13. individual adj. 个性的
14. formulate vt. 构想出

通用词汇

① evolutionary adj. 进化的
② diversity n. 多样性
③ graphite n. 石墨
④ spontaneous adj. 自发的

8.3 Sketchbooks

Sketchbooks are the repository of a fashion designer's ideas, observations and thoughts. Whilst there is no template[15] for the perfect[16] sketchbook (and they are not solely the preserve of fashion designers), a good fashion sketchbook should enable the designer to progressively record and document a series of ideas and inspirations through related visual and written material accumulated[17] over time.

All sketchbooks evolve in response to changing influences and circumstances. The true value of a sketchbook is in how the designer uses it to pause and reflect on their work in a meaningful way in order to continue to the next stage of the design journey. It can sometimes be difficult to fully comprehend this when starting out; there may be a temptation to fill up the opening pages with lots of secondary images but this will not lead to a personal sketchbook unless it starts to take on the personality of the user, rather like a personal diary or journal. A sketchbook should become as individual as your fingerprint and provide you with a growing resource from which ideas and concepts can be explored[18] and developed without feeling self-conscious. Sketchbooks also enable you to explore and develop your own drawing style; the book will build up over time and its resource value will increase. One of the most useful aspects[19] of a sketchbook is its portable[20] nature, allowing you to carry it around and enter quick thumbnail sketches or observational drawings.

Most fashion student sketchbooks are A4-size. However there is no fixed rule on this as some students successfully work with A3-size sketchbooks. Sometimes working across a landscape⑤ A3 format can be useful for sketching A4-size fashion figures and developing preliminary line-ups. The smaller A5 pocket-size sketchbooks can be useful for discreetly⑥ carrying around; they also work well as fabric swatch[21] books and for entering additional thumbnail sketches.

8.4 The fashion figure

The proportions of a fashion figure are often exaggerated[22] and stylized, particularly for women's wear drawings. This can sometimes be slightly confusing to the untrained eye but in fashion terms it represents a statement of an ideal rather than an actual body shape. This ideal is then aligned to a contemporary look that is viewed through the visual lens of fashion.

8.3 速写本

速写本是时尚设计师理念、观察和思维的储藏室。但是没有完美速写本的模板（没有针对时尚设计师携带的速写本），一本好的时尚速写本应该是设计师不断累积图形和文字材料，记录一系列理念和灵感的档案库。

速写本受环境影响不断变化。速写本的真正价值在于设计师如何用这种有意义的方式停下来认真思考他们的工作，以便继续下一阶段的设计。刚开始时有时很难完全理解这种方式，有一种诱惑，要用很多二手图片资料填满空白页面，如果这样，速写本就没有个性，除非开始时它就具有使用者的个性特征，而不是像个人的日记或者个人的行程计划。速写本应该像人的指纹那样具有个性，给你提供越来越多的资源，使你能够无意识地从中研究和发展理念和概念。速写本也能使你研究和发展你自身的绘画风格，随着时间的建构，速写本的资源价值也在不断地增加。速写本最大的好处是携带方便，不管去什么地方你都可以带着它，快速地画一些大拇指大小的草图和观察到的事物。

时尚学生的速写本大多数是 A4 大小。但是没有统一的标准，就像有些学生使用 A3 尺寸的速写本很顺手，有时画风景用的 A3 尺寸速写本画 A4 尺寸的时尚人像和加注配文时很有用。口袋大小、A5 尺寸的速写本也是非常有用的，能够很体面地到处携带，它也可以当作面料小样本和画一些拇指大小的草图。

8.4 时尚人像

时尚人像经常比例夸张和具有风格，特别是画女装。有时可能误以为眼光没有经过训练，但是在时尚领域它表达一种理想，而不是现实中人体的体型。这种理想与当代形象相一致，是时尚透视镜里的形象。

专业词汇

15. template n. 模板
16. perfect adj. 完美
17. accumulate vt. 逐渐增加
18. explore vt. 探索
19. aspect n. 方面
20. portable adj. 便携式
21. swatch n. 样本
22. exaggerate vt. 夸张

通用词汇

⑤ landscape n. 风景
⑥ discreet adj. 谨慎的

Since the late 1960s and 1970s exaggerated proportions have generally prevailed[23] and continue to exert an artistic influence over most fashion drawings. Most standing fashion figures are proportioned between nine and ten heads in height. Most of the additional height is gained through the legs[24], with some added to the neck[25] and a little added to the torso[26] above the natural waist[27]. Most women in the real world stand around 5ft 5in or 5ft 6in, but a fashion figure needs to project greater height in order to better show off the clothes and communicate the look to an audience, usually through exaggerated gestural poses[28]. Of course, a woman who might be 5ft 2in could be proportioned the same as a woman standing 5ft 2in but for fashion purposes neither would offer the desired ideal proportions for communicating the look. When drawing the fashion figure the look might refer to the prevailing styles of the season, such as the position of the fashion waist, or it may be an exploration of voluminous or contoured[29] clothing styles with reference[30] to influences from particularly favoured[31] model or celebrity.

There are fundamental differences between the fashion proportions for drawing men and women. Women's fashion proportions are mostly concerned with extending height through the legs and neck, with the resulting drawings taking on a sinuous⑦ and gently curved appearance. For men the drawing approach is altogether more angular[32].

8.5 Drawing from life

Drawing from life, which is an excellent way to develop and refine your drawing skills, involves observational drawing of real-life male or female figures. It is important to consider the appropriate art materials and media, such as charcoal⑧, pen or pencil, as well as paper type[33] and the eventual scale[34] of work. Drawing is a process that can be improved[35] and enhanced with regular[36] practice and life drawing offers the particular opportunity of developing and improving hand-to-eye coordination. This is essentially about trusting yourself to spend more time looking at the figure in front of you, rather than by glancing[37] at the figure then looking at the emerging drawing itself and drawing from memory.

This is a common mistake among life-drawing students. It is very important to study the figure before you start to draw. Try to make sure that you are in a good viewing position and then analyze the pose. If the figure is standing, it is essential to establish which leg is taking most or all of the weight[38]; this will critically determine the stability[39] of the pose in relation to what is called the "balance[40] line". The balance line is an imaginary line that drops from the base of the centre of the neck down to the floor at the position of the foot[41]. It can be drawn on the paper and used as a guide to ensure that the figure remains standing without "tipping over" on the page. As a general rule, the leg that is supporting the weight of the pose, which should always be drawn before the other leg, will curve down to the floor and should join up with the balance line at the outside edge[42] of the foot.

自从20世纪60年代后期和20世纪70年代大多数时尚绘画逐渐流行夸张的比例，并受艺术的影响，大多数站立的时尚人像高度都在9～10个头之间的比例。增加的高度大多数分配在腿部，还有一些在颈部，另外一点点增加在自然腰以上的身躯部位。现实中大多数女性的高度大概是1.4～1.42m之间，但时尚人像要塑造得高些，并用夸张的姿势，为了更好地展示服装，与观众交流款式。当然，一位1.32m的女性也可以和一位1.52m女性具有相同的比例，但她们两者的理想比例不足以交流时髦样式。在画时尚人像时，样式应该是这个季节流行的样式，例如时尚腰线的位置，或者是受某个知名模特儿或明星的影响，对服装的体积、轮廓线的研究。

男性和女性的人像绘画比例有根本性的差别。女性时尚比例通过延长腿部和增加颈部高度，呈现出蜿蜒柔美的画面，而男性时装画强调棱角。

8.5 临摹生活

临摹生活是发展和增强你绘画技能的最好方法，包括对真实男性和女性人像的临摹。重要的是考虑用适当的艺术材料和媒介，例如墨石、钢笔或铅笔，不同种类的纸和纸的大小。通过不断地实践和临摹生活，在其过程中，绘画的能力得到提高，同时也是培养手眼协调的最佳机会。花较多的时间注视你面前的人像画出来的画，与匆匆一瞥、凭记忆画出来的画相比，使你更加相信自己。

学生在人像写生时普遍存在错误。在开始画之前研究人体非常重要。你设法使自己处在一个视角好的位置上，然后分析姿势。如果人像是站立的，那么就要确定哪条腿承受大部分或全部重量，它对决定姿态的稳定性很关键，与平衡线相关。平衡线是一条想象的线条，从脖颈底部中心，到地板上脚的位置。它可以画在纸上，作为一条引导线，确保人像保持站立，不至于倾倒翻出纸面。一般规则是，姿势中支撑重量的那条腿，要比另一条腿先画，从上到下呈现弧线，平衡线落在脚的周围。

专业词汇

23. prevail vi. 盛行
24. leg n. 腿部
25. neck n. 颈部
26. torso n. 躯干
27. waist n. 腰部
28. pose n. 姿势
29. contour vt. 画轮廓
30. reference n. 参照
31. favour vt. 喜爱
32. angular adj. 角度
33. type n. 类型
34. scale n. 规模
35. improve vt. 改进
36. regular adj. 有规律的
37. glance vi. 一瞥
38. weight n. 体重
39. stability n. 稳定
40. balance n. 平衡
41. foot n. 脚
42. outside edge 边缘线

通用词汇

⑦ sinuous adj. 蜿蜒的
⑧ charcoal n. 木炭

The principle of the balance line applies to all standing fashion poses including those simulating a walking pose. It is also applicable to menswear although men's poses are generally made less dramatic[43] and gestural than for women's fashion drawing.

Studying the pose first also allows time to evaluate[44] distinctions between the "actual" figure and the expression[45] of an "ideal", fashion figure for womenswear or menswear. Proportions in fashion drawing represent an ideal, so it follows that the life figure does not need to be drawn as an exact representation[46]. This requires interpretative[47] visualization, which is an essential release[48] for fashion drawing.

8.6 Creating poses

Fashion drawings are frequently[49] characterized by gesture[50] and movement[51], both of which are ideally suited to exploration through drawing the fashion figure from life. Part of a fashion drawing's allure is its seemingly effortless style, which is sometimes the result of a careful selection of lines and what is left to the imagination of the viewer. In this regard it is important to note the value of line quality in the fashion-drawing process. Line quality describes the varieties of drawn lines or marks that have their own inherent characteristics depending on the media that is used, the paper quality, the speed at which the line is made and even the angle of the pen or pencil as it moves along the surface[52] of the paper. Distinct from adding tone and shading techniques, the use of line to convey essential information is integral[53] to most fashion drawings.

Some of the most expressive and visually engaging fashion poses are the result of linear drawings, where selective line quality is used to maximum[54] effect. An understanding of fashion proportions and the standing balance line is essential as a building block[55] for more gestural poses, which instill movement and personality into a fashion drawing. In addition to studying poses from life, it is also possible to develop poses by tracing over figurative photographs in magazines, but this needs to be approached with care: consider the image only as a starting point. Fashion is, after all, a human activity[56] so it follows that developing and creating studied poses is a useful exercise and will aid the development of templates or sketch for future use.

It should be possible plate figure to different design ideas. While the pose should be relevant[57] to the context[58] of the clothing (for example, it would make little sense to draw a sporty pose for a wedding dress or an evening gown), creating the pose is much more about the body underneath. Look for movement lines that run through the body—not the outline[59] of the figure—noting the intersections[60] of the pose at the bust[61], waist and hip[62] positions. The leg supporting the weight must be grounded, but the other limbs[63] can be modified or adapted to enhance gestural qualities. In this way, the resulting fashion poses can exaggerate the "actual" to project a more expressive "ideal".

平衡线原则适用于所有站立姿势，包括那些模拟行走的姿势。在表现动态的男装时尚画也要考虑平衡线，但是时装画中男性姿势的动态没有女性的夸张，姿态的变化也不多。

无论是画女装还是男装，首先要花一些时间研究姿势，评估真实人像和理想的时尚人像之间的差别。在时尚绘画中比例代表了一种理想，因此不需要精确表现真实形象，需要的是可解释性的视觉效果，这就是时尚绘画表现的关键所在。

8.6 创造人体姿势

时尚画通常以姿势和动态为特征，两者要从时尚画的写生中探索。一些时尚画的魅力看上去很轻松，实际上线条经过精心计划，留给观者去想象。从这方面来看，在时尚绘画过程中线的特征有很重要的价值。线的特征就是绘画中线的变化，或者是使用不同媒介体现出的内在特征。例如纸张的质量、画线的速度，甚至钢笔或铅笔在纸表面上移动时的角度等。与增添调子和阴影技巧不同的是，线条表达了本质信息，大多数时尚绘画都少不了线条。

一些最具表现力和最具视觉吸引力的姿势，就是线描的结果。通过选择不同特征的线条，取得了最佳的效果。为了画出更多的动态，了解时尚比例和站姿平衡线是很重要的，并且将它们渗透到时尚画的动感和个性特征中。此外，从生活中和杂志上的人像照片都能研究姿态，但要谨慎对待：研究图像只是出发点。毕竟，时尚是人的一种活动，只有不断地研究和创新姿态，才是有效的锻炼，有利于发展时尚画模板或速写，为将来所用。

一种姿势可以不止一次使用，一种人像模板可以表现不同的设计理念。但是姿势应该与服装的内容一致（例如，采用一种运动性姿势表现婚纱或晚装是毫无意义的），创新姿势时要更多地考虑内在的身体。寻找贯穿身体的活动线条——不是身体的轮廓线，注意胸部、腰部和臀部位置横断面的姿势。支撑身体重量的腿必须着地，另外一条腿用来修饰，增强姿势的美感。用这种方式可以夸张时尚姿势，将实际形象表现为更加理想的形象。

专业词汇

43. dramatic adj. 戏剧性
44. evaluate vt. 评估
45. expression n. 表达
46. representation n. 表现
47. interpretative adj. 解释
48. release n. 释放
49. frequent adj. 频繁的
50. gesture n. 姿势
51. movement n. 运动
52. surface n. 表面
53. integral adj. 完整体
54. maximum adj. 最大值的
55. block n. 块面
56. activity n. 活动
57. relevant adj. 相关的
58. context n. 上下文
59. outline n. 轮廓线
60. intersection n. 交叉点
61. bust n. 胸
62. hip n. 臀
63. limb n. 肢体

8.7 Heads, faces and hair

Fashion heads, facial features and hairstyles are worthy of special consideration in fashion drawing; they can convey a multitude of essential style and gender[64] information.

The very personal and unique attributes[65] that a face can contribute to a drawing are worth exploring through practice and exercises. Much like the evolution of fashion drawing itself, the "ideal" face changes over time and takes on many guises. Make-up trends continue to have a direct influence on contemporary fashion faces and it is always useful to collect magazine tear sheets from which to study and evaluate different faces and proportions.

Although faces can be drawn in the linear style that is so often used in fashion, they can also lend themselves to applications of tone and shade. Structurally, the forward-facing head is oval in shape for women-much like an egg shape-and should be horizontally[66] intersected at mid-point[67] to position the eyes. The mouth[68] is usually arranged halfway[69] between the eyes and the base of the chin. The mouth could be considered in two parts with its upper and lower lips[70]. The upper lip should include an "M" shape definition[71]. The nose[72] may either be represented with dots for the nostrils[73] above the top lip of the mouth or with an added off-centre[74] vertical[75] line from the front of the face as if to indicate a shadow[76]. Noses are rarely given any prominence[77] in fashion faces as the eyes and lips become the main features.

The ears may be discreetly added at the side of the head starting at eye level[78] and ending just above the nostrils; they can be useful for displaying earrings[79], if appropriate. The hair should be carefully considered as this can have a transforming effect on the appearance of the fashion head. Again, collect tear sheets from magazines in order to build up a visual file of hairstyles as it can be quite challenging to imagine them without a reference point and of course, hairstyles for women vary enormously. If it is visible the hairline[80] should be drawn around a quarter of the way down from the top of the oval shape of the head. Line, shade and color can all be added according to the style requirements.

8.8 Arms, hands, legs and feet

When drawing a fashion figure it is important to consider the hands, arms, legs and feet in relation to the pose and gestural qualities. When drawing an arm, consider it in three parts: the upper arm[81], the elbow[82] and the lower arm[83]. The upper arm is attached to the shoulder[84] from which it may pivot[85] depending on the angle of the torso. It has a smooth, gently tapering[86] upper section[87] that reaches down to the elbow position. The lower arm tapers more visibly to where it joins the hand.

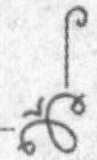

8.7 头、脸和头发

在时尚画中，时尚的头部、脸部特征和发型值得特别关注。它们可以表达多种基本风格和性别信息。

通过实践和练习，探索脸部的个性化和独特特征是非常值得的。就像时尚画自身的变化那样，理想脸形也是随着时间而变化的，并采用不同的化妆。化妆潮流不断地对当代时尚的脸形产生直接影响。将从杂志上撕下的脸形小纸片汇集在一起，便于研究和评估各种脸型和比例。

时尚画中经常用线画脸部，还可以用色调和阴影来润饰它。从结构上来说，女性正面头型是椭圆形，很像鸡蛋的形状，眼睛在水平中点位置上，嘴在眼睛和下颌底部的中间位置，嘴由上嘴唇和下嘴唇两部分构成，上嘴唇呈“M”形状。鼻子在上嘴唇上方，用点表示鼻孔，或者在脸部中间增加一条偏离中心的垂直线来表示阴影。时尚画中，很少突出鼻子，而是突出眼睛和嘴唇。

在头的两侧谨慎地添加耳朵，上端与眼睛平齐，下端对齐鼻孔上方。如果需要的话，画上耳环是很有用的，头发应该仔细考虑，因为它能够改变时尚头型外貌。同样，从杂志上收集发型图片，可以建立发型的视觉档案，因为没有参考资料的想象很艰辛，当然，女性发型千变万化。可见的发际线位置从椭圆形头顶向下 1/4 处。根据风格添加线条、阴影和色彩。

8.8 手臂、手、腿和脚

在画时装人像时，还要考虑手、手臂、腿和脚，与姿态特征的关系。在画手臂时，将手臂分成三个部分：上臂、肘部和前臂。上臂与肩部相连，根据躯干的角度可以旋转。从上部分向下到肘部逐渐变细，前臂到手的部位变得更细。

专业词汇

64. gender n. 性别
65. attribute n. 特征
66. horizontally adv. 水平地
67. mid-point n. 中点
68. mouth n. 嘴
69. halfway adv. 中间
70. lip n. 嘴唇
71. definition n. 定义
72. nose n. 鼻子
73. nostril n. 鼻孔
74. off-centre 偏离中心
75. vertical adj. 垂直的
76. shadow n. 阴影
77. prominence n. 突出
78. level n. 水平
79. earring n. 耳环
80. hairline n. 发际线
81. upper arm 上臂
82. elbow n. 肘部
83. lower arm 前臂
84. shoulder n. 肩部
85. pivot vi. 枢轴
86. taper vt. 逐渐变细
87. section n. 部分

The hands have two main parts: the front or back of the palm[88] and the fingers[89] and thumb[90]. Both parts may be elongated[91] to offer the fashion figure a range of gestures and actions, which will all enhance the drawing. Consider the angle of the lower arm when drawing the hand. Fingernails may be included but knuckles are not usually emphasised[92]: too much detail on a hand can make it look wrinkled[93]. You could also try drawing the hand resting on the hip with the fingers hidden from view.

The feet are usually drawn in a simplified way that mostly assumes a shoe[94] line. When starting out it is helpful to practice sketching bare feet, but the foot will usually be hidden from view within a shoe, which can be drawn in a huge variety of styles. The overall look will be determined by the angle of the foot and whether or not the shoe has a heel.

As fashion drawing is largely concerned with presenting an interpretation of an ideal figure rather than realistic proportions, so it follows that drawing the legs is an exercise in artistic licence. Fashion legs are routinely extended in the upper leg and thigh, and below the knee[95] to where the ankle[96] meets the foot.

手主要由两部分构成：掌心、手背和四指、拇指。这两个部位都可以拉长，为时尚人像提供各种姿势和动作，为此增加绘画的效果。画手时，要考虑前臂的角度。可能要画指甲，但不要强调指关节，手上太多的细节使手看起来布满了皱纹。也可以尝试着将手画成搭在臀部上，手指隐藏起来看不见。

脚经常使用简单的方式来画，主要画鞋子的线条。开始学画时尚画时，画光脚是非常有用的。但是，脚通常藏在鞋子里，鞋子可以画成很多种样式。脚的角度决定鞋子的整体形状，决定是否画鞋子的后跟。

因为时尚画主要关注理想形象的表现形式，而不是真实的比例，因此画腿部时要从艺术的角度来画。按常规，时尚的腿在腿上部和大腿处、膝盖下方到脚都要延长。

专业词汇

88. palm n. 手掌
89. finger n. 手指
90. thumb n. 拇指
91. elongate v. 延长
92. emphasise vt. 强调
93. wrinkle vi. 起皱纹
94. shoe n. 鞋
95. knee n. 膝盖
96. ankle n. 脚踝

Exercises

(1) Understanding the text.

Read the text and answer the following questions.

1) What kind of fashion drawing do you like?

2) What are the differences between fashion drawing and traditonal drawing?

3) What's the application of fashion drawing in fashion design?

(2) Building your language.

The following expressions can be used to talk about fashion drawing. Choose the right ones to fill in the blanks in the following sentences. Change the form where necessary.

leading	ready-to-wear	fabric	original
incorporate into	glamorous	original	elegance

This is an ____________ artwork for a fashion illustration by Marcel Fromenti for The Lady, a weekly magazine for women published since 1885. At the time it was made, Fromenti was the main artist for The Lady's fashion articles. The ____________ women in his drawings modelled both couture and high-end ready-to-wear garments with equal panache and ____________. Couture dresses and suits by ____________ Paris and London couturiers such as Christian Dior, Pierre Balmain and Norman Hartnell were drawn with the same flair as designs from British ____________ labels such as Susan Small, Roecliff & Chapman, and Marcus. The articles described the fashion developments of their day in simple, ____________ terms that contributed greatly to The Lady's popularity with its readers. Pencil notes record the designers, ____________ and colour details, alongside technical instructions to the printers as to how these images should be ____________ the printed page and at what scale.

(3) Sharing your ideas.

After learning about fashion drawing in this section, are you eager to share your knowledge? Please write a short introduction (of around 300 words) to introduce fashion drawing. Try to make full use of what you've learned from text, including the relevant information from the reading text as well as words and expressions.

It is important for designers to understand
as early as possible how a garment grows
from a two dimensional concept into a three-
dimensional object. A pattern is a flat paper
or card template, from which the parts of the
garment are transferred to fabric [illegible]
cut out and assembled.
A good understanding of body shape and how
body measurements transfer to the pattern
piece is essential. The pattern cutter2 must
work accurately[1] in order to ensure that
once constructed, the parts of fabric fit together
properly and precise[illegible]
A block (also [illegible] sloper3) is a two-
dimensional te[illegible] garment form
(for example, a [illegible] or fitted skirt) that
can be modified [illegible] elaborate[2] design.
Blocks are cons[illegible]ing measurements
taken from a size[illegible] a live model, and
do not show any sty[illegible]nes or seam allowance.
Blocks must, howe[illegible] include basic amounts
of allowance for ease6 and comfort; for instance,
a tight-fitting bodice block would not have as much

Unit 9

Fashion Cutting

第9单元

时尚裁剪

It is important for designers to understand as early as possible how a garment grows from a two-dimensional concept into a three-dimensional object. A pattern is a flat paper or card template, from which the parts of the garment are transferred[1] to fabric, before being cut out and assembled.

A good understanding of body shape and how body measurements transfer to the pattern piece is essential. The pattern cutter[2] must work accurately① in order to ensure that once constructed, the parts of fabric fit together properly and precisely.

9.1 The block

A block (also known as asloper[3]) is a two-dimensional template for a basic garment form (for example, a bodice[4] shape or fitted skirt) that can be modified into a more elaborate② design. Blocks are constructed using measurements taken from a size chart[5] or a live model, and do not show any style lines or seam allowance. Blocks must, however, include basic amounts of allowance for ease[6] and comfort; for instance, a tight-fitting bodice block would not have as much allowance added into the construction as a block for an outerwear garment might. A fitted bodice block would also have darts added into the draft[7] to shape the garment to the waist and bust, whereas a block for a loose-fitting overcoat[8] would not need these.

9.2 The pattern

A pattern is developed from a design sketch using a block. The designer or pattern cutter will add to the block by introducing style lines, drapes, pleats[9], pockets and other adjustments[10] to create an original pattern.

The final pattern features a series of different shaped pieces of paper that are traced[11] on to fabric and then cut out, before being seamed together to create a three-dimensional garment.

Each pattern piece contains notches[12] or points[13] that correspond to a point on the adjoining pattern piece, enabling whoever is making the garment to join the seams together accurately. The pieces need to fit together precisely, otherwise the garment will not look right when sewn together and it will not fit well on the body.

When the block modification is finished, seam allowance is added to the pattern. To perfect a pattern, a toile (a garment made out of a cheap fabric such as calico) is made and fitted on to a live fitting model. Adjustments can be made on the toile before being transferred to the pattern.

9.3 Pattern cutting

Like all craft skills, pattern cutting can at first seem difficult and intimidating. But with a basic understanding of the rules to be followed (and broken!), the aspiring designer will soon learn interesting, challenging and creative approaches to pattern cutting. To draw the right style line in the correct position on a garment takes experience and practice. Designers who have been cutting patterns for twenty years can still learn something new—the process of learning never stops. This makes creative pattern cutting a fascinating process.

设计师较早地知道一件服装如何从二维概念变成三维的物体，是非常重要的。服装纸样就是纸或硬纸的平面模板，再将纸样转移到面料上，然后裁剪和缝合。

重要的是要对体形有深刻的了解，并且知道如何将测量的尺寸转移到纸样的衣片上。纸样裁剪师必须仔细地工作，确保缝制时服装衣片完美、精确地缝合在一起。

9.1 原型

原型（也称为服装尺寸样本）是基础服装的二维样板模型（例如贴身上衣或贴体裙子），在其基础上变化成更加复杂的设计。原型采用的尺寸来自尺寸表格或真实人体，没有任何风格线或者缝份。但是原型必须包括人体活动和舒适的基本量。例如，上身贴体原型的松量就没有外衣的上身原型松量多。贴合上身原型要增加省道，形成腰和胸部形状，而宽松服装就不需要这些省道。

9.2 纸样

纸样是由设计草图和原型发展而来。设计师或者样板师在原型上增加款式线垂褶、褶裥、口袋和其他修改，设计一个最初的纸样。

最终纸样有一系列不同形状的衣片纸样，将它们转移到面料上，裁剪和缝合，得到一件三维的服装。

每个样片上要打剪口或标记点，和其他样片连接点相对应，不管是谁制作服装，都能将衣片精确地缝合到一起。衣片必须精确地缝合在一起；否则，当服装缝制好以后会看上去不对劲，也不能很好地贴合身体。

当原型修改完成后，纸样上要添加缝份。为了得到完美的纸样，需要用坯布（用便宜的面料例如白坯布）制作服装，再到人体上试穿。试穿后，先在坯布上进行修改，再转移到纸样上。

9.3 纸样裁剪

像所有工艺技能那样，起初纸样裁剪似乎很困难，很令人畏难。但是，当理解（打破）基本规则以后，有追求的设计师将很快学会用有趣的、挑战性的、创造性的方法去进行纸样裁剪。要想在服装的正确位置上放置准确的风格线需要经验和实践。已有20年裁剪经验的设计师仍然需要学习新知识，学习过程从未停止，创新样板就成为令人着迷的过程。

专业词汇

1. transfer vt. 转移
2. pattern cutter 样板师
3. sloper n. 原型
4. bodice n. 上衣身
5. size chart 尺寸表
6. ease n. 松量
7. draft vt. 画纸样
8. overcoat n. 大衣
9. pleat n. 褶裥
10. adjustment n. 调整
11. trace vt. 拓印
12. notch n. 刻痕
13. point n. 点

通用词汇

① accurate adj. 精确的

② elaborate adj. 精心制作的

9.4 Samples[14]

A sample is the first version of a garment made in real fabric. It is this garment that goes on the catwalk or into a press/showroom[15]. Samples are produced for womenswear in sizes 8 ~ 10 to fit the models.

Once the sale book is closed, the samples are stored in the company's archive[16]. Some samples of past collections are taken out by designers for photo shoots, events such as premieres③ and for reference or possible inspiration for future collections.

9.5 How the measurements relate to the block

Whether taking individual measurements or using a size chart, the main measurements (bust girth[17], waist girth, waist-to-hip length and hip girth) will give a good indication of the body shape the design is intended to fit. Secondary measurements may also be taken from an individual or from a size chart.

Darts can be used to control excess[18] fabric and to create shape on a garment when stitched[19] together.

9.6 How to read a design drawing

This is the point at which pattern cutting becomes much more creative and exciting. Once the design has been completed, the process of breathing life into a flat design drawing in order to achieve an actual garment can begin. To be able to achieve a beautiful garment shape takes time and experience. Remember nothing ever happens without practicing your skills—don't be disheartened④ if it doesn't work first time round. All outstanding fashion designers and creative pattern cutters have worked for years to perfect their skills.

9.7 Translating drawing to block

The translation of a design drawing to pattern requires an eye trained for proportions. Most design drawings are sketched on a figure with distorted proportions. The legs and neck are too long and the figure too slender[20]. These sketches are often inspiring and wonderful to look at but unfortunately give a false image of the human body and it is a key task of the pattern cutter to address this.

9.4 样衣

样衣就是用真实面料制作的第一件服装样品。正是这件样衣要拿到T台、媒体和展示厅展示。女装样衣以8~10型号制作。

一旦销售订单结束后，样衣被放进公司的档案。一些过去收藏的样衣被设计师拿出来拍照，或参加一些首映式等活动，或作为未来作品的参考资料和灵感来源。

9.5 如何测量原型的相关尺寸

无论是从个体量取尺寸，还是使用尺寸表的尺寸，一些主要尺寸（胸围、腰围、腰臀的长度和臀围）能很好地指示身体的体型，并能为此体型设计贴体的服装。其他尺寸可能来源于个体或尺寸表。

省道可以用来控制多余的面料，当面料缝合起来时，可以设计服装的廓形。

9.6 如何阅读设计画稿

从这点上看，纸样裁剪有很大的创造性和激动性。一旦设计完成后，将平面设计画稿变得有生气，直到变成一件真实服装的过程就开始了。为了获得一件漂亮样式的服装，需要耗费时间和经验。记住没有实践技巧，任何事情不会发生——如果第一次作品不完美，不要灰心。所有优秀的时尚设计师和纸样创意裁剪师都是工作了很多年以后才使他们的技能得到完善。

9.7 设计转变成原型

将一种设计画稿转变成纸样需要对比例训练有素的眼光。大多数设计画稿的人像比例都是变形的。腿部和颈部太长，人像太纤细。这些草图通常看上去令人激动和完美，但是不幸的是提供了虚假的人体形象。对于纸样裁剪师来说，必须纠正这些错觉的东西。

专业词汇

14. sample n. 样衣
15. showroom n. 陈列室
16. archive n. 档案
17. girth n. 周长
18. excess n. 多余量
19. stitch n. 缝迹线
20. slender adj. 苗条

通用词汇

③ premiere n. 首次公演
④ disheartened adj. 沮丧的

9.8 How to mark the block

It is essential when cutting a block or a pattern that the correct information is supplied. A bodice block, for example, has to show the horizontal lines of the bust, waist and hiplines[21]. Parts of the block such as the waist and bust points should be notched or punch[22] marked (holes and notches indicate where the separate pieces of fabric will be attached[23] to one another) and the grain line[24] must be indicated. This will clearly show the position in which the pattern should be placed on the fabric. Additional in formation must be written clearly in the centre of the block, including whether it is a front or back piece, a tight-or loose-fitted bodice block and the sample size, preferably with the measurements and any allowances to be made when constructing the block.

Once the pattern has been constructed the seam allowance can be added. Seam allowance can vary in size from a narrow 0.5cm for a neckline[25] (to avoid having to clip or trim the seam) to 2.5cm in the centre back of trousers[26] (to be able to let some out if the waist gets too tight). Seams that are to be joined together should always be the same width. Mark the width of the seam allowance on the block.

Usually, the block ends up being divided into further pattern pieces. At this point, therefore, the information should be reconsidered accordingly, except the grain line and front or back information, which is always transferred to the new pieces.

9.9 Dart manipulation

Darts control excess fabric to create shape on a garment. They can be stitched together end to end or to a zero point also known as the pivotal point (such as the bust point). Dart manipulation is the most creative and flexible part of pattern cutting. The possibilities are endless and the designer's imagination is the only limitation. Darts can be turned into pleats, gathers[27] or style lines. Their positioning on the body is very important; not only do these techniques create fit, shape and volume, they also change the style and design of the garment.

9.10 Slash[28] and spread[29]

This method is used to add extra volume and flare[30]. The technique involves creating slash lines that reach from one end of the pattern to the other, sometimes ending on a pivotal point like a dart ending. These slash lines will then be opened up for added volume and flare.

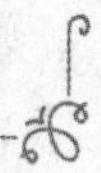

9.8 如何标记原型

当裁剪原型或者纸样时，提供正确信息是很重要的。例如，上身原型必须显示胸部、腰部和臀部的水平线。原型的某些部位，例如腰线和胸点，要有凹痕或打孔的标记（孔和凹痕表明衣片在这里对齐缝合）。纱线的方向必须表示，这将清晰地表示纸样放置在面料上的位置。其他信息必须清楚地写在原型的中央位置，包括衣片或后衣片，贴体或宽松的上身原型和样板尺寸，更周到的是标上尺寸和制作原型时的放松量。

一旦纸样完成后，就要加放缝份。不同部位缝份的量不同，在颈部窄一些，约0.5cm（避免打剪口或修剪缝份），裤子的后中线需要2.5cm（如果裤腰太紧，可以放出来一些）。缝制的时候缝份要一样宽，在原型上标记缝份的量。

在原型基础上分割发展的纸样，信息要作相应调整，除了纱向和前后衣片信息，所有其他的信息都要转移到新的纸样上。

9.9 操作省道

省道能控制服装上多余的面料，从而创造形。它们可以从一端缝制到另一端，或者从一端到零点，也叫做关键点（例如胸点）。省道的操作是纸样裁剪上最具创造性和灵活性的。省道有无穷多可能性，而设计师的想象极其有限。省道可以变成褶裥、抽褶或风格线，这些手法应用在身体的何处很重要，可以设计贴体、塑造形状和体积，也能改变服装的样式和设计。

9.10 剪口和展开

这种方法用来增加额外和展开的量。这种技术包括设计一条剪切线，从纸样的端到另一端，有时在一个关键点结束，像省道的端点；然后拉开这些剪切线，增加量和展开。

专业词汇

21. hipline n. 臀围线
22. punch vt. 打孔
23. attach vt. 附着
24. grain line 纱向线
25. neckline n. 领围线
26. trousers n. 裤子
27. gather n. 碎褶
28. slash vt. 切口
29. spread vt. 展开
30. fare vt. 张开

9.11 Sleeves[31]

Sleeve construction is a very special part of pattern cutting. Sleeves can be part of the bodice (laid-on sleeve) or set into an armhole (set-in[32] sleeve). Without any other design features added, a garment can look outstanding by simply creating an interesting sleeve design. The most basic sleeve block is the one-piece[33] (set-in) sleeve. Different sleeve blocks can be developed from the one-piece block, such as the two-piece[34] sleeve and laid-on sleeves, including raglan[35], kimono/batwing[36] and dolman[37] designs.

When constructing a set-in sleeve, the measurement of the armhole is essential. Therefore, the bodice front and back are constructed first and once the measurement of the armhole is established, ease is added according to the type of block (jacket block, fitted bodice block and so on). Ease is added to a pattern to allow for extra comfort or movement.

As well as allowing the sleeve to sit comfortably in the armhole, ease will also affect the fit and silhouette of a garment. Ease is distributed between the front notch and the double back notch of the sleeve. In some set-in sleeve designs, the ease is taken across the shoulder to achieve a round appearance over the shoulder point. A sleeve is sitting comfortably in the armhole when it aligns exactly with, or is set slightly in front of, the side seam of the bodice.

There are differences between one-piece and two-piece sleeves, the major one being the amount of seams that are used. A one-piece sleeve has only one seam placed under the arm at the side seam position. Therefore, the seam cannot be seen when the arm is relaxed. The two-piece sleeve has two seams; one is placed at the back, running from the position of the back double notch down to the wrist, past the elbow. The second seam is moved a little to the front, from under the arm side seam position (still not visible from the front). The look of a two-piece sleeve is more shapely and it has a slight bend[38] to the front. As such, it is possible to get a closer fit with a two-piece sleeve because of its extra seam. One-piece sleeves are used for a more casual look, whereas two-piece sleeves are mostly seen on garments such as tailored jackets or coats[39].

The laid-on sleeve is part of the bodice. Once constructed, either a part of the armhole remains or there is no armhole at all. A laid-on sleeve is most commonly constructed by separating the one-piece sleeve through the shoulder notch straight[40] down to the wristline[41] to gain a front piece and a back piece. The next step is to align the front piece of the sleeve with the bodice's front shoulder and the back sleeve with the bodice's back shoulder. From this point onwards several styles can be developed, such as batwing or kimono, raglan, gusset and dolman sleeves. The sleeve can be laid on at variant angles—the greater the angle, the more excess fabric and therefore a greater range of arm movement.

9.12 Collars

The collar is a versatile design feature that will enhance the style of a garment. It is attached to the neckline of the garment and allows the size and shape of the neckline to vary. Collars come in all shapes and sizes and the most common are the stand-up[42]/mandarin[43], shirt, flat, sailor[44] and lapel collar[45] constructions.

9.11 袖子

袖子结构是纸样裁剪中相当特别的部分。袖子可以是上衣身的一部分（套袖）或者安装在袖窿里（装袖）。一件服装不需要增加任何其他特征，只创造一个有趣的袖子，就能使服装看上去非同一般。最基本的袖子原型是一片袖（装袖）。不同袖子原型可以从一片袖发展而来，包括两片袖和连肩袖，例如和服袖和蝙蝠袖。

装袖结构、袖窿尺寸很重要。因此，设计衣身的前衣片和后衣片袖窿结构，然后测量袖窿，要根据原型的不同类型增加放松量（夹克的原型、贴体的上衣身，等等）。加放的松量应保证穿着的舒适性和运动量。

袖子放松量不仅可使袖子充裕地安装在袖窿里，还影响服装的贴体性和廓型。放松量均匀分布在袖子前记号缺口到后记号缺口之间。一些装袖设计中，袖子放松量安排在肩部，使肩部有圆润的效果。如果袖子安装得准确，袖子充裕地安装在袖窿里，或者从上身侧缝处略向前倾。

一片袖和两片袖有很多不同，主要不同之处在于它们的分割缝不同。一片袖只有一条缝，在手臂下方侧缝的位置。因此，当手臂下垂时看不到缝。两片袖有两条缝，一条缝在后面，从衣片袖窿的缺口标记处向下通过肘部到手腕；第二条缝从手臂下方侧缝的位置稍稍向前偏移（从前面看不到缝）。两片袖子的样式更加有形少许向前弯曲。正是多了一条缝，两片袖有可能更加贴体。一片袖多用于休闲服装，而两片袖大多用于合体的夹克或外套。

套肩袖是上衣身的一部分。它的构造要么保留一部分袖窿，要么没有袖窿。最普通的套肩袖沿肩线到手腕有一条分割线，分成前袖片和后袖片。接下来就是将前袖片与上衣身前片的肩缝合，后袖片与上衣身后片的肩缝合。依此进一步发展为多种样式，例如蝙蝠袖或和服袖、套袖、三角插片袖、肩袖。袖子安装的角度有多种——角度越大，需要面料越多，因此，手臂的运动范围也越大。

9.12 领子

领子可以设计成多种风格，能够增强服装的风格。它与服装的领线缝合在一起，领线的尺寸和形状可以变化。领子有各种形状和大小，最常见的领子结构是站领或中式领、衬衫领、水手领和驳头领。

专业词汇

31. sleeve n. 袖子
32. set-in 装袖
33. one-piece 一片
34. two-iecc 两片
35. raglan n. 插肩袖
36. batwing n. 蝙蝠袖
37. dolman n. 斗篷
38. bend n. 弯曲
39. coat n. 外套
40. straight adj. 直的
41. waistline n. 腕围线
42. stand-up 立领
43. mandarin n. 中式领
44. sailor n. 水手
45. lapel collar 驳头领

Collars can be constructed in three basic ways. The first method is a right-angle[46] construction, used for stand-up collars, shirt collars[47] and small flat collars[48] such as Peter Pan and Eton collars. Secondly by joining the shoulders of the front and back bodice together to construct the collar directly on top of the bodice block. This technique is used to construct sailor collars and bigger versions of flat collars. Finally, the lapel construction, which is extended from the centre front, from the breaking point toward the shoulder. By extending the break/roll line[49] a collar construction can be added. A version of this is the shawl collar, where the collar extends from the fabric of the garment on to the lapel without being sewn on.

领子有三种基本构成方法。第一种方法是直角结构，有站领、衬衫领和小平领，例如彼得潘领和伊顿领。第二种方法是将前后衣片的肩部缝合起来，然后直接将领子安装在原型衣身的上端。水手领和大型平坦的领子就是这种构造。第三种领子是驳头领结构，沿前中心线延伸，从翻折点向肩部翻折。通过延伸折线或翻折线，增加领子结构。其中之一是披肩领，领子从服装的衣片向外延伸到驳头，没有缝合线。

专业词汇

46. right-angle 直角
47. shirt collar 衬衫领
48. flat collar 平翻领
49. roll line 翻领线

Exercises

(1) Understanding the text.

Read the text and answer the following questions.

1) Whatare the methods of fashion cutting?

2) Please try to understand the evolution of men's style from the angel of cutting.

3) What is clothing design innovation in our country?

(2) Building your language.

Translate the following sentences into Chinese.

1) Consumer preference for environmentally safe products has lead some manufacturers to look for fabrics made with naturally colored cotton.

____

2) American designer score high when it comes to marketing savvy and making salable clothes that appeal to the whole U. S. population.

____

3) Due to the technology advances and the wide use of synthetics, less than 10 percent of all buttons sold in the United States are constructed of natural materials.

____

4) Shoe companies may have shoe lines in all price ranges and categories.

____

5) His treatment stabilizes the fabric so that there will be no further change in its size or shape and therefore improves the fabric's resilience.

____

(3) Sharing your ideas.

After learning about fashion cutting in this section, are you eager to share your knowledge? Please write a short introduction (of around 300 words) about fashion cutting. Try to make full use of what you've learned from Text, including the relevant information from the reading text as well as words and expressions.

Unit 10
Fashion Show

第10单元
时尚秀

A catwalk fashion show is a sales promotion
mechanism in the clothing industry recognized
cultural event. Although the fashion show is
essential to how the fashion industry works, it
has also become a cultural icon in its own right.
A fashion show is a biannual presentation
of a new clothing collection on moving bodies
for an audience. A new collection is produced
by a designer, brand company, or group of
companies. The parade of moving bodies
makes up an essential feature of a fashion
show, and [illegible] given [illegible] to the modeling
profession as [illegible] a [illegible] of conventions of
movements [illegible] accompanied
by music [illegible] the rhythm of
movement [illegible] ands from
the overall [illegible] oving bodies are
predominant[illegible] ale [illegible] ough menswear
fashion show[illegible] since the late
1929s they are [illegible] numbered by
women's wear [illegible]
A fashion sho[illegible] and its modes of presentation
may be explained to a large extent in terms of

A catwalk fashion show is a sales promotion[1] mechanism in the clothing industry recognized cultural event. Although the fashion show is essential to how the fashion industry works, it has also become a cultural icon in its own right.

A fashion show is a biannual presentation of a new clothing collection on moving bodies for an audience. A new collection is produced by a designer, brand company, or group of companies. The parade of moving bodies makes up an essential feature of a fashion show, and has given rise to the modeling profession as well as to a range of conventions of movements, poses and looks. It is accompanied by music which emphasizes the rhythm of movement and blocks out other sounds from the overall impression. The moving bodies are predominantly female. Although menswear fashion shows have been held since the late 1929s, they are still by far outnumbered by women's wear show.

10.1 Framing the fashion show

A fashion show and its modes of presentation may be explained to a large extent in terms of frame analysis. This applies both to the spatial (setting, catwalk, set and runway design), and temporal (music, performance, staged appearances) framing of the fashion (See Figure 10.1). Framing devices include the technologies, props and conventions that set the fashion show apart from ordinary interaction and define what is going itself both within the fashion show itself and between the fashion show and the outside world.

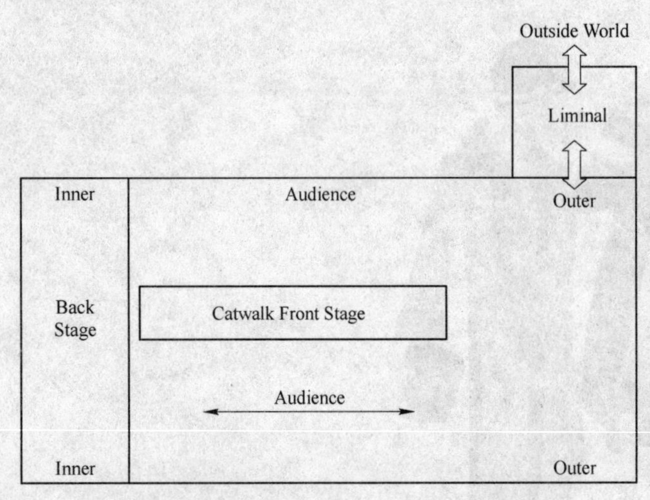

Figure 10.1　The fashion show framework

Firstly, fashion shows are set apart from the outside world in terms of their location. As part of a Fashion Week programme, fashion shows are often held in conjunction with trade fairs in exhibition grounds that are typically (but not always) located on the outskirts of large and medium-sized cities. The atmosphere in such locations (whether they are exhibition hall or marquee tent) is neutral and anonymous[2]. Typically, they have no windows and the fact that they are totally enclosed enables the staging of the fashion show to be completely controlled. In this way, the attention of the invited audience is directed away from the outside world and made to focus entirely on the ephemeral[3] setting that frames the fashion show performance.

T台时尚秀是服装行业的促销手段，也被广泛认为是文化活动。尽管时尚秀本质上是展示时尚企业的工作，但它自身已经具有文化的象征含义。

时尚秀就是通过模特儿的移动，为观众展示一年两次的服装新品。新的服装产品由设计师、品牌公司或公司集团生产。活动的模特儿展示，构成了时尚秀的重要特征，因此产生了模特儿职业和一系列约定俗成的走步方式、姿势和外貌。为了强调运动的节奏感和免受其他声音的干扰，配有音乐，构成整体印象。模特儿主要是女性。尽管从1929年后期已经举行男性的时尚展示，但它仍然不及女性服装的时尚秀多。

10.1 时尚秀的组织结构

时尚秀和它的展示模式很大程度上可以用结构分析来解释。时尚秀采用了空间性（背景、T台、座位和台设计）和瞬时性（音乐、表演、舞台形象）这两种结构（见图10.1）。结构设置包括技术、道具和一些常规做法，这就使得时尚秀与正常的互动活动不同，定义了正在举行的时尚秀的内在内容，也将时尚秀与外部世界隔离开来。

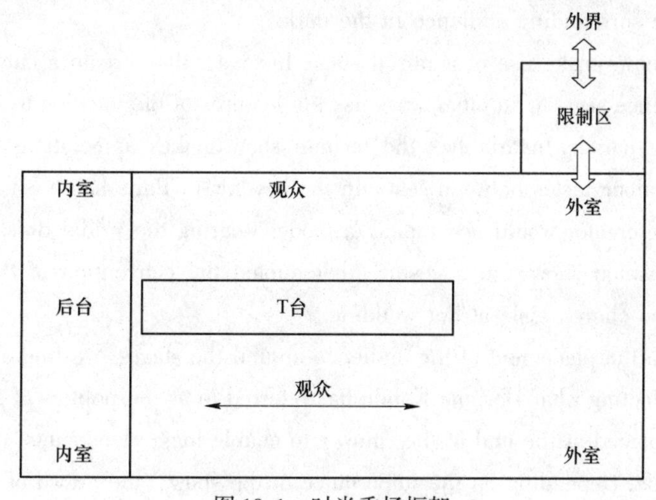

图10.1 时尚秀场框架

首先是举办时尚秀的地点与外界隔开。作为时尚周的部分日程，时尚秀通常在展览馆的贸易展销会上举行，一般（但不总是）在大型和中等城市的郊区。在这些地方（展厅或大帐篷）的气氛一般没有倾向性和特色。通常没有窗户，事实上它们是完全封闭的，时尚秀的表演完全在控制之中。用这种方法，受邀观众的注意力远离外部世界，完全聚焦于时尚秀转瞬即逝的表演情景中。

专业词汇

1. promotion n. 促销
2. anonymous adj. 匿名
3. ephemeral adj. 短暂的

In addition to this type of neutral setting, fashion shows are also held in locations that are chosen to color the atmosphere of the show. In French and Italian fashion shows aristocratic ancient régime palais may be selected, while other typical location include derelict factories, warehouses[4], theatres and museums. In reality, therefore, fashion shows are held in all sorts of locations.

Secondly, the importance of the presentation is marked on a vertical plane by the procession of models. The parade typically takes place on a raised stage. In the golden age of haute couture, the stage was referred to as the podium. Since then, other terms have taken over, including catwalk, signifying a narrow passage, and runway, which—with a reference to the takeoff of an airplane—refers to the launching of a new collection. The raised dais—like a theatre stage, college high table or church altar—gives ritual significance[5] to the activities performed, and exalts the persons performing, there, thus separating the audience from the performers, those who look from those who are looked at. The direction of gazes is re-enforced by lighting that bathes the runway in strong light and leaves the surrounding audience in the dark.

Not all fashion shows make use of a raised stage. Instead, they create a catwalk by making an aisle between audience seats or in other ways use the features of the location to create a space visibly laid out for the parade. Invariably, the fashion show makes associations to other situations where people walk along aisles between seats. In the heyday[6] of Paris haute couture, at the end of the parade the male creator would accompany a model wearing the bridal dress, traditionally the last number in a fashion show, in a gesture that quoted the convention of the father leading a young woman up the church aisle at her wedding.

When it comes to the placement of the audience around the stage, we find a whole set of framing conventions reflecting what Dorinne Kondo has referred to as the politics of seating. Invariably, photographers are placed at the end of the runway to enable long—lens shots[7] of the models walking down the catwalk. Depending on the importance of the show, the crowd of photographers may vary from a few to a veritable forest of telephoto lenses and cameramen, although for TV, webcast transmissions or videotaping, two to three cameras give the best coverage. This means that space needs to be available for not only a head-on spot, but also a side view and a position closer to the start of the runway for the "return" shot. Photographers also need to have access to positions from which to shoot the guests, especially the "dignitaries" in the front-row[8]—if not during the show itself, then immediately before or after it. So photographers are given privileged visual positions underlining the importance of the mediation of the event to audiences not present.

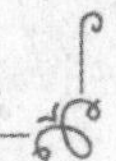

除了在没有倾向性的地点之外，时尚秀也选择在气氛丰富多彩的地点举行。在法国和意大利也许选择在贵族气息的古老宫殿里举行时尚秀，其他一些典型的地点包括废弃的工厂、仓库、剧院和博物馆。事实上时尚秀可以在各种地点举行。

其次重要的是模特儿行走的垂直跑道，跑道往往高出地面。在高级时尚的黄金年代，跑道称为表演台。从那以后，其他一些术语就取代了这个名词，包括猫步舞台，即一条窄的通道和跑道，引自飞机起飞的跑道，意味着新系列服装的推出。升起的舞台——像剧院的舞台、学院高的讲台或者教堂的圣坛——使表演活动富有仪式的含义。表演者高度的提升，与观众区分开来，表演者与观众彼此被对方观看。T台上沐浴着强烈的光线，光照加强了凝视的方向，使T台周围的观众处在黑暗中。

并不是所有的时尚秀都采用高出地面的舞台，也可以把观众座位排在两边，中间形成一个甬道成为T台，其他方法就是使用地点特征，创造一个可以走步的视觉空间。不变的是，时尚秀总是在两排座位之间设置可以走步的甬道。在巴黎高级时尚鼎盛时期，时尚秀最后压台的是一位男模陪伴一位身穿新娘婚纱的女模走到T台的末端，引用了新娘父亲搀着新娘在教堂甬道上行走的传统形式。

当谈到舞台周围观众的位置时，我们发现了一整套惯常的组织结构，Dorinne Kond 称之为政治化座位。毋庸置疑，摄影师被安排在T台的末端，使他能够用长焦镜头拍摄正在T台行走的模特。摄影师的人数由时尚秀的重要程度而定，或许只有几个，或许像森林般的长焦镜头和摄影人。虽然是为电视、广播的传输或录像拍摄，只有两到三台相机有好的视角，这意味着要留有正前方、两侧的空间供拍摄，以及T台开始的位置，拍摄模特返回的镜头。摄影师还需要有一个能拍到贵宾的位置，特别是指定坐在第一排的那些名流。如果不在秀场里，也要在秀场开始前或秀场结束后赶紧拍摄他们。因此，摄影师给予视角位置特权，强调了不在秀场的观众对时尚秀的重要影响力。

专业词汇

4. warehouse n. 仓库
5. significance n. 意义
6. heyday n. 盛世
7. shot vt. 拍摄
8. font-row 前排

In the politics of seating, choice spots are determined by their proximity① to, and view of, the action on the catwalk. The seating area in front of the cameras at the end of the stage is considered to provide the best view when available. In general, though, the best seats at a fashion show are in the front row at the end of or along the stage. These front row seats are reserved for the most important guests, such as magazine editors, who are the essential filters through which the shows are reported in the media, and celebrities, whose presence may add prestige to the show. In sales shows, buyers are also seated in the front row, but today most buyers will view the collection informally in the showroom, so that the purpose of the fashion show is increasingly to present an overall[9] image for the press[10], and only indirectly for the buyers. The seats behind the first row are for less important guests including many buyers and business contacts, company employees, design school student and other members of the public interested in attending the fashion show. In large fashion shows there may be a standing area behind the VIP seats. In other shows, the first row is extended by manipulating the space, so that everyone in the audience can have a first row seat. This is possible in fashion shows held in large premises where the catwalk area can be extended, sometimes through several rooms and corridors. It can also be done if the parade of models trails around or through audience seating arrangements.

At the back of the catwalk is the set design, which serves as the backdrop of the performance. A fashion show is typically accompanied by a set of slides, projecting the logo[11] and credits[12], as well as images, colors and designs that enhance the concept of the show. The set design also separates front stage, where collections are appreciated and consumed by the audience, from back stage, where they are pieced together and produced concerned. As such it marks the point where models change their staged pace as they prepare to leave or enter the front stage theatre. While the front stage is carefully scripted in its staged framing devices, both in place and time, in order to exclude all possibility of unscripted behavior and individual improvisation in the ritual performed, the back stage consists of ordered chaos—order in the necessary arrangement of clothes enabling models to hurriedly dress, change and dress again, but chaos in the sheer number of different kinds of personnel present and the multiplicity② of tasks that they must carry out to enable the front stage performance to take place.

In this framework, the fashion show can be said to consist of two performances encased in each other. The first one starts with the arrival of the audience, which is obliged to form a queue to enter a single access point to the fashion show stage (often via a liminal space between the outside world and the show venue), and every member of which is vetted and passed or rejected by gatekeepers who examine printed invitations and check individual names as printed on their invitation lists. The start of the show is almost invariably delayed, which incidentally gives everyone time to observe the crowd and spot which editors and celebs grace the show with their presence. VIP guests may calculate the delay and time their arrival at the venue accordingly, with the more famous being allowed to arrive later than the hoi polloi.

在政治化座位中，座位安排由T台上活动的远近程度和视野决定。在正对T台、照相机所处位置的前排座位有最好视野。一般来说，时尚秀中最好的座位是T台末端和T台两侧的第一排座位。这些前排座位预留给最重要的贵宾。如杂志编辑，媒体报道的秀场信息要经过他们的筛选；明星，他们的出席为秀赢得了声誉。在销售性的时尚秀中，买家也坐在前排，但是，今天大多数买手都是在非正式的展示厅里观看新品系列。因此，越来越多的时尚秀是对媒体展示完整的形象，只是间接为买家。前排后面的位置安排不太重要的宾客，包括很多买家和生意往来者，公司员工、设计学校的学生和其他一些对时尚秀感兴趣的公众。大型时尚秀在VIP座位后面还有站位。另一些时尚秀，通过空间设置延伸前排座位，使每位观众都是前排座位。如果在一些大型场所举办时尚秀，有可能将T台区域延伸，有时穿过几个房间和走廊。同样，如果模特沿着或穿过观众的座位行走，效果也是一样。

T台后面是背景设计，衬托表演。时尚秀通常配一组幻灯片，包括logo和企业的荣誉，图像、色彩和设计，强调时尚秀的概念。背景设计将舞台分为前台和后台。前台是欣赏和购买发布会产品的观众；后台是根据设计师的意图将单件服装组合并放置在一起。背景也是模特准备离开后台或进入前台的标志点，在这点上，模特开始改变为舞台步伐。前台台面从步伐和时间上经过精心设计，为了正式表演中排除所有未计划的和个人的即兴行为。后台有序而混乱，服装必须有序摆放，才能保证模特急匆匆地穿、脱和更换服装。混乱是因为后台有各类人员，他们各自的任务不同，确保前台的表演顺利进行。

在这种框架结构中，可以说时尚秀由两个互相包含的表演部分构成。第一部分是由观众的到达开始，他们被要求排队从单一的通道进入会场（通常很小的空间连接外部与秀场），每一个人都要被检查，门卫检查印刷的请柬，核对姓名是否在他们邀请的名单上，决定放行还是拒绝。节目开始的时间总要被推迟，正好留出时间每个人可以观看其他人，目睹编辑和名流们的出席给秀的增色。VIP贵宾可能计算延迟的时间和他们到达的时间，越是出名的人越是可以比普通人晚到。

专业词汇

9. overall adj. 全部的　　11. logo n. 标志
10. press n. 新闻报道　　12. credit n. 信誉

通用词汇

① proximity n. 接近　　② multiplicity n. 多样性

The second performance, the performance of the models on stage, starts with the outbreak of music—usually so loud that it drowns all other sounds together with an adjustment of lighting. It is at this point that the first model appears on stage. The music accompanying a fashion show is selected and played by a DJ in order to match the designer's concept for the show. Together music, lighting and slides are used to emphasize discrete sections in the collection presented. The fashion show usually lasts for no longer than fifteen to twenty minutes. Its end is signified by the appearance of all the models who parade together down the runway to the accompaniment of the audience's applause. Eventually, the designer whose collection has been shown also makes an appearance sometimes brief and informal, sometimes obviously choreographed③. Not infrequently, a few members of the audience will come up to the catwalk to hand a bouquet of flowers to the designer. After this, the fashion show has ended and the audience leaves. For many, fashion shows are part of a busy fashion week schedule[13], so they may well be rushing on to the next appointment.

10.2 Backstage production

Backstage, a large number of people work to realize the show. A relatively basic fashion show involves around twenty people—excluding models and support personnel such as caterers and drivers—and can easily run to a budget of €60000. By comparison, for top designer shows, such as those by Dior or Chanel, figures of five million dollars are quoted. In spite of the variations, which do occur, there are bundles of tasks and lines of command that are common. They make for a routinized④ interaction which is necessary for the success of an event that is usually produced under considerable time pressure.

In fact, preparations start well in advance of the fashion show. Typically, a designer or the fashion house concerned approaches an event agency[14] six months before the planned show to talk about concepts and budgets. The event maker or art director of the event agency presents a concept, which is perhaps modified, but otherwise accepted by the company, and the event maker will then start the actual preparations for the show. In the proposal, some of those who will be involved in the production are named—for example, the stylist, an interior decorator, possibly a photographer to document the event, the production manager, persons in charge of lights and sound, and possibly one or two top models. These people will have been approached in advance and asked to join the project[15]. Upon acceptance, the professionals discuss and come to agreement on the proposal, often supplementing details in their own area of expertise[16]. The involved parties will then prepare their own part in the production, and the art director will present additions or changes to the fashion house for approval. Next a venue is chosen. Normally, it has to be coherent with the theme of the show, unless the latter takes place in a venue set up to house different shows—for example, a tent connected with a fashion fair[17]. If this is the case, the following account will need to be modified since lighting, sound, decorations and so on will for the most part already have been put up.

第二部分表演是舞台上模特儿的表演，随着音乐突然响起——声音之大淹没了其他声音——同时调整灯光。正在此时，第一个模特儿出现在舞台上。时尚秀的配乐由DJ挑选和演奏，要与设计师的概念相吻合。音乐、灯光和幻灯片加在一起，使发布会的离散部分得到强调。发布会持续的时间在15～20min。所有模特儿一齐出现在T台上，表示秀结束，并伴随观众的掌声。最后发布会的设计师亮相，有些人简短而不正式，有些人明显设计了舞蹈动作。通常有些观众到T台前给设计师送花。在这之后，时尚秀结束，观众离开。对很多人来说，时尚秀是时尚周繁忙日程的一部分，因此，他们也许会匆忙地赶到下一个预约地点。

10.2 后台工作

后台有很多人为秀的举行而工作。相对简单的秀大概需要20人——包括模特儿和一些协助人员，例如饮食提供者和驾驶员——经费预算大约需要6万欧元。与高级设计师的秀相比，例如Dior或Chanel，需要500万美元预算。尽管这种差异存在，但是任务的多少和逐层控制程序是相同的。要制定合作路线图，在时间紧迫的压力下，这是活动取得成功必须要做的。

事实上，在举行时尚秀很久之前准备工作就已经开始了。通常设计师或时尚屋在举办秀场的前6个月就开始与代理接洽，讨论方案和预算。代理公司的制作人和艺术指导提出一种方案，公司或许提出修改意见，或许同意，然后制作人就开始为秀进行实际的准备工作。在这种方案下，要邀请一些人参与，例如，造型师、室内装饰师，可能的话要请一名摄影师要为此项活动做一些档案的拍摄工作，产品经理、灯光和音响负责人，或许还有一、二名超级模特。事先和这些人商量，邀请他们参加这个项目。接受邀请之后，他们开始专业性的讨论，对提案取得一致意见，通常各自从自身专业角度提出细节方面的补充意见。然后，这个团队开始准备自身的那部分工作。若有补充或变化，艺术指导要得到时尚屋的认可。接下来就是选择场地。一般来说，地点要与发布会的主题相一致，除非是在一个举行多场秀的地点——例如时尚贸易会设立的帐篷。在这种情况下，要做的事情就是修改已经安装好了的灯光、音响、装饰等等。

专业词汇

13. schedule n. 时刻表
14. agency n. 代理
15. project n. 项目
16. expertise n. 专家
17. fair n. 商品交易会

通用词汇

③ choreograph vt. 设计舞蹈动作　④ routinize vt. 使惯例化

The backstage venue consists of separate stations where the different professionals have their base. This is in order to organize the somewhat chaotic ad hoc workspace, and to make communication easier. Hair and make-up are located in a faraway corner; stylist at the entrance to the runway; people in charge of sound, lighting and decoration all around the front area (see Figure 10.2). Most professionals have assistants, and throughout the day, different people arrive and begin their separate jobs. As it is not necessary for the models or only when the hairdresser, make-up artist, and stylist are ready to start preparing them. Similarly, waiters and seaters will not be present until shortly before the invitees arrive. Depending on the type of fashion house and the scale of its show, the designer or designers will not be at the venue until shortly before the guests.

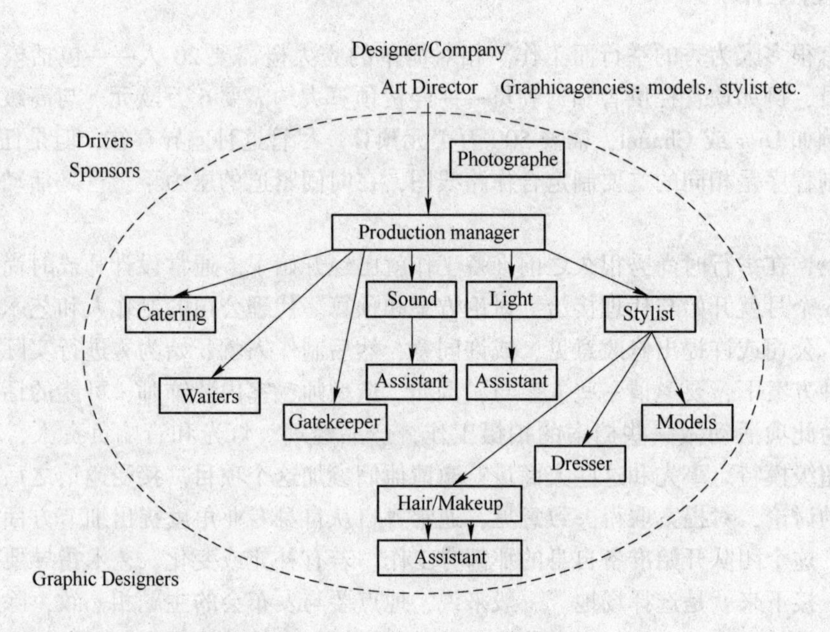

Figure 10.2 Fashion show organization

Throughout the day, rehearsals[18] are carried out. This is mostly to estimate the time and make sure those models, choreographer, lighting and sound technicians know what to do when. Simultaneously, clothing might be stitched up or ironed[19], lights may be put up, and seating arranged. Dressers receive their instructions and, as each dresser often dresses up more than one model, and each change must be done in a few seconds, all the clothes are hung up unzipped and unbuttoned in exact order. The models are introduced to their dressers, and during the event, they will go from the stage to the dressers and wait for them to finish dressing the previous model. To help organize outfits, a photograph of each model is put up at each wardrobe station. At big shows there is a table of order of outfits that serves as a visual reference for all details.

后台分为几个不同的专业性区域，以使得有点混乱的临时工作场所条理化，以便彼此的沟通容易些。发型师和化妆师在后面角落里；造型师在T台的入口处；负责音响、灯光和装饰的人在前面区域（见图10.2）。大多数专业人士都有助手，不同的人在一天中不同的时间到达，准备他们各自的工作。例如，安装灯光和椅子的时候不需要模特和穿衣工。模特儿是在发型师、化妆师和造型师开始为他们做准备时才到达。类似地，服务者和座位引导员只是比客人早到一会儿。不同的时尚屋和时尚秀的规模设计师或设计师们比客人早到一会儿。

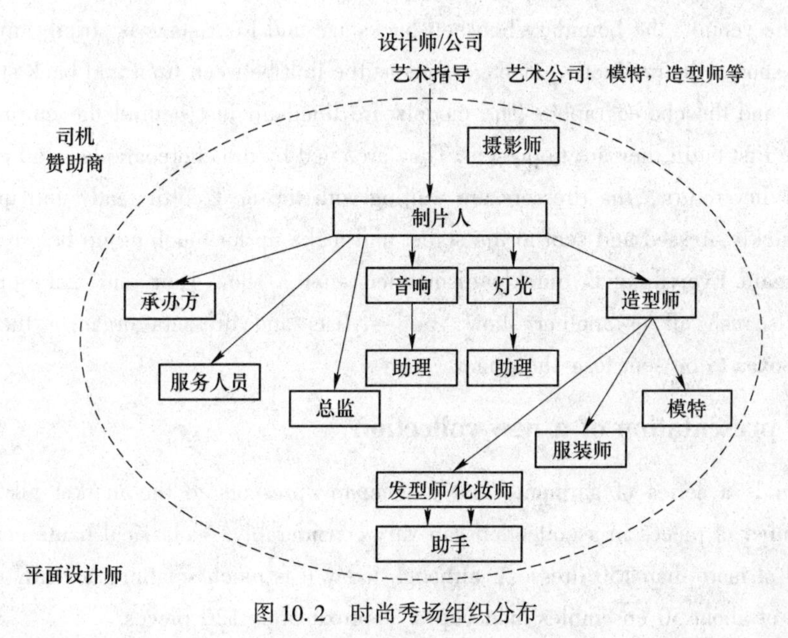

图10.2　时尚秀场组织分布

在白天要进行排练。排练后才能计算准确的时间，使模特儿、编舞者、灯光和音响技术员知道什么时候该做什么。同时，服装需要缝合或熨烫，灯光要测试，座位要排放，穿衣工要接受讲解，因为每一个穿衣工要为多个模特儿穿衣，每套服装在几秒钟之内更换完成。所有服装敞开拉链和扣子，按次序准确地悬挂在衣架上。模特儿与他们的穿衣工见面，在发布会的时候，他们从T台走向他们的穿衣工，等候前面模特儿穿好衣服后为他们更换服装。为了便于服装的搭配，每个模特儿的衣服站点都张贴了此模特事先拍好的一组照片。大规模的秀场，有一整套服装搭配的次序表，可以根据图片参考所有细节。

专业词汇

18. rehearsal n. 排练专家　　19. iron vt. 熨烫

By the time the show is ready to begin, the backstage area will be filled with people who are all connected to the realization of the event. As the venue is often small, only people who are essential are allowed into the backstage area. The production manager is now very important since the level of concentration in each area makes him or her the only link between the different professionals, and the only one who has an overview of the entire production. Hair and make-up are now finished, and are on standby for touch ups during the show. The people in charge of music and lighting are present in the front stage area to keep an eye on the show while it is in progress. As the guests enter the venue, the boundary between backstage and front stage is strictly upheld.

During the show, the production manager acts as the link between front and back stage, cueing lights, sound and the choreographer. The models are lined up just behind the curtain[20] waiting, dressed in the first outfit they are to present. They are cued by the choreographer and enter the catwalk, and as they return, the dressers are waiting with the next outfit ready and unzipped. The models are quickly dressed and sent to the stylist and make up for touching up before going out on the catwalk again. Everything is quickly dismantled after a show. Hair and make-up pack their things, models rush off to another show, and stylists and dressers organize the clothes on hangers[21] in boxes to be sent to a showroom.

10.3 The presentation of a new collection

A collection is a series of garments that a company presents to the market all at the same time. The number of pieces in a collection can vary considerably. A classical haute couture collection consisted of more than 150 dresses, although today it is much smaller; a ready to wear show often consists of about 50 ensembles made up of approximately 120 pieces.

There are a number of fashion shows that do not present a single brand or designer, but a group thereof. Group shows include graduation shows, staged by most design schools where each graduate will present a small collection, typically consisting of three to eight outfits. Similarly, fashion show contests, usually for young designers, consist of multiple small collections, produced under conditions specified by the content organizers. Apart from the entertainment[22], the purpose of such shows is not to sell clothes, but to showcase the capabilities[23] of individual designers, both for the press and for potential employers. In addition, fashion fairs often open with a trend show, which presents a selection of garments by the exhibitors at the fair.

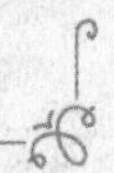

秀即将开始时，后台挤满了为秀服务的人。因为后台很小，只有那些关键人物允许进入后台。现在产品经理相当重要，因为每个工作站点集中于自己的工作，他或她成了唯一连接各个站点的人，也是唯一能够看到整个秀场的人。发型师和化妆师已经完成工作，站在一旁等待演出时的补妆。负责音乐和灯光的人在舞台前面不断注视秀的进展。当贵宾进入秀场后，后台和前台被严格控制。

在秀进行的时候，产品经理起着联结前台和后台的作用，要给灯光、音响和编舞者发出信号。模特身穿要展示的第一套服装排在幕布帘的后面等候出场。他们在编舞者的暗示后进入 T 台，当他们返回时，穿衣工已经等在那里，为他们脱掉服装，换上准备好的下一套服装。模特儿很快地换好服装、送到造型师那里，再次进入 T 台前要补妆。秀结束后，所有东西很快被撤走。发型师和化妆师打包他们的东西；模特儿匆忙离开去另一个发布会；造型师和穿衣工将服装用衣架整理好，或放进箱子里，然后服装被运送到展示厅。

10.3 发布会新系列服装

一个系列就是公司在同一时间呈现给市场的一组服装。发布会的服装件数有很大差别。典型的高级时尚发布会至少 150 件套，也许今天数量上少很多；成衣发布会至少 50 套，即至少约 120 件。

有很多时尚秀不仅仅展示单个品牌或单个设计师，而是很多人的作品。这样的秀包括很多设计学院举行的毕业秀，即每个毕业生展现数量少的系列作品，通常 3 ~ 8 套。类似地，为了竞赛举行的秀有很多年轻设计师参加，风格各异，每个系列数量较少，是根据主办方的要求进行的设计。这种秀的目的不是为了销售服装，而是为了娱乐、对媒体或想招聘新人的企业展示设计师的个性能力。此外，时尚交易会经常举行趋势发布会，其服装由交易会承办方在交易会的摊位上选取。

专业词汇

20. curtain n. 窗帘
21. hanger n. 衣架
22. entertainment n. 娱乐
23. capability n. 才能

10.4 On living bodies

Whether a show is an haute couture, ready-to-wear, group, or trend show, the clothes are presented on living bodies.

As moving images became an integral part of fashion show presentations, a strange interplay of movement and montage was created. The staccato⑤ movements of the models suggested a montage of still images, and a certain number of end poses were in play as models rested momentarily at the end of the runway, offering opportunities for the perfect image for photographers. The movements rather resembled those of soldiers on parade but while the movements of soldiers are choreographed to be firm, determined and very abrupt[24], the catwalk movement for especially female models had to be soft, swaying[25], and spherical, to reflect particular ideals. Thus, the upper body is kept erect[26] and passive[27], with the arms dangling[28] carelessly along the sides while maybe holding on to a bag[29]. The knees are lifted higher than in a normal walk, with each leg swayed exaggeratedly over the other as the moving foot is placed in front of the one behind. All in all, this movement pattern causes the model to look like an idealized or stylized object[30].

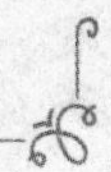

10.4 活动的人体

不管是高级时装秀、成衣秀，群组秀，还是趋势秀，服装都采用活动的人体展示。

由于活动的形象是时尚秀展示的关键，一些奇怪的相互作用的动作和蒙太奇形式被设计出来。模特的一些不连贯动作，就像一系列静态图形的蒙太奇，在T台的末端模特摆出很多姿势，以便摄影师有机会拍摄完美的图片。这种运动很像士兵被检阅时的步伐，但是士兵的步伐被设计成坚定的、坚决的和相当突兀的。而T台的步伐，特别是女性模特，必须是柔软的、前后摇摆的和球状的，表现一种特定的理想。因此，上身保持直立和被动，手臂在两侧无意地摇摆，也许拎着一个包。膝盖比正常走路时抬得要高，一条腿夸张地摇晃着跨到另一条腿前面，移动的脚落在另一只脚的前面。总之，这种运动的图式使模特看上去像一个理想化或风格化的物体。

专业词汇

24. abrupt adj. 唐突的
25. sway vt. 摇摆
26. erect adj. 直立的
27. passive adj. 被动的
28. dangle vi. 悬荡
29. bag n. 包
30. object n. 物体

通用词汇

⑤ staccato adj. 不连贯的

Exercises

(1) Understanding the text.

Read the text and answer the following questions.

1) How can we design a good fashion show?

2) How can we display diversities of art in fashion show ?

3) What are the problems in fashion show in China?

(2) Building your language.

The following expressions can be used to talk about fashion show. Choose the right ones to fill in the blanks in the following sentences. Change the form where necessary.

integrate	combine with	enterprises	textile
promotion	anniversary	consumption	fiber

On March 12th, on the third ____________ of the start of green fiber logo certification, China Chemical Fibers Association held the first "Green Life, Green Fibers-Green Fibers Theme Fashion Show" to further strengthen the ____________ of green fibers logo certification, further ____________ the scientific and technological achievements of green development with industrial channels, and promote the green process of chemical fibers industry. At the same time, ____________ the dynamic and static display of the green ____________ logo products in many chapters such as dress, home, baby, sports and health, more ____________, users and consumers can understand the green fibers logo products, thus showing the vivid practice of green fibers in the green process of ____________ industry in recent years and the bright vision of future development, so as to make "Green life, Green fibers". The ____________ concept is deeply rooted in people's hearts.

(3) Sharing your ideas.

After learning about fashion show in this section, are you eager to share your knowledge? Please write a short introduction (of around 300 words) about fashion show. Try to make full use of what you've learned from Text, including the relevant information from the reading text as well as words and expressions.

Unit 11

Draping

第11单元

立体裁剪

Draping1 method is improved by the industrial development through 1800's to twentieth century. The invention of tape measure2, invention of sewing machine in 1851 in 1863 of the development of paper patterns and in 1911 the production of muslin padded mannequins are the very important developments that occurred in the fashion industry. By the effect of these improvements couture designers started to use three dimensional models to drape the fabric and check3 the relation betwe[illegible] the design and the body.

1. Fashion [illegible] flat pattern making

There are [illegible] that can be taken when [illegible] into a 3 D object for the first [illegible] approach that is chosen is very much [illegible] by the design that needs t be achieved, a[illegible] particular working styles of the design[illegible] or the pattern maker.

The first approach i[illegible] flat pattern making whereby a pattern [illegible] aker will look at the designers sketch and will select a basic pattern

Draping[1] method is improved by the industrial development through 1800's to twentieth century. The invention of tape measure[2], invention of sewing machine in 1851, in 1863 of the development of paper patterns and in 1911 the production of muslin padded mannequins are the very important developments that occurred in the fashion industry. By the effect of these improvements couture designers started to use three-dimensional models to drape the fabric and check[3] the relation between the design and the body.

11.1 Fashion draping and flat pattern making

There are two main approaches that can be taken when converting a design into a 3D object for the first time. The approach that is chosen is very much decided by the design that needs to be achieved, and the particular working styles of the designer and/or the pattern maker.

The first approach is by flat pattern making, whereby a pattern maker will look at the designers sketch and will select a basic pattern block from which to build the design, making alterations to a basic jacket pattern for instance and adding fullness[4], moving seamlines, changing necklines etc, until the pattern resembles the design. Then this pattern can be made up in a test fabric and a first fitting can be done on a dress makers dummy[5] or mannequin or on what is known as a fit model—a live model with the exact body measurements of the pattern. After these changes will be made, the pattern will be altered and a new toile recut until the designer and pattern maker are happy that the design is correct. This approach is predominantly based around flat pattern making, and is called "flat" because it is predominantly done to the 2D pattern on the pattern making table using existing patterns and measurements.

The second approach is very different and is much more about what the fabric wants to do, or what the fabric will allow you to do. This approach is known as "drape" or "working on the stand". Fashion draping is an important part of fashion design. Draping for fashion design is the process of positioning and pinning fabric on a dress form to develop the structure of a garment design. A garment can be draped using a design sketch as a basis, or a fashion designer can play with the way fabric falls to create new designs at the start of the apparel design process. A designer or pattern maker will begin by draping basic fabric, such as calico[6], onto a mannequin and working much more like a sculptor the fabric is smoothed, creased[7], spliced[8] and pinned until the desired shape is achieved. Once the garment is roughly the right shape, the fabric can then be marked with pen lines and notch marks and annotations[9] of what piece is which, so that it can then be removed from the mannequin and flattened[10] out without later confusion. It is from this drape that a first pattern can then be traced and the lines and measurements smoothed[11] and checked before a new first toile is sewn and ready for a fitting on a fit model.

从19世纪到20世纪的工业发展，立体裁剪技术得到了改进。1851年卷尺和缝纫机的发明，1863年纸样的发展、1911年衬垫的平纹细布人体模型产品的出现，它们都对时尚行业的发展起到了重要作用。正是由于这些方面的改进，高级时装设计师开始使用三维模型立体裁剪面料，检验设计和身体之间的关系。

11.1 时尚立体裁剪和平面纸样制作

将最初的一种设计转变为3D物体，主要有两种方法可以采用。究竟采用哪种方法，一种是由设计决定，即需要取得什么样的效果；另一种是由设计师或样板师独特的工作方式决定。

第一种方法是平面纸样制作，样板师根据设计师的草图，选择一种基本原型样板，以它为基础，根据设计再做适当修改。例如，一件夹克基本的样板，可以增加丰满度，去除接合线，改变领线等等，直到样板与设计相似。然后，这种样板采用试验的面料，第一次在人台或人体模型试穿，或者模特试穿一种真人模特进行试穿，其身材与样板尺寸比较吻合。试穿之后将做一些改动，在纸样上做修改，重新用坯布裁剪，直到设计师或样板师认为正确体现设计为止。这种方法采用平面样板制作，称为平面是因为二维样板根据已有样板和尺寸在样板桌上完成。

第二种方法截然不同，更多关注希望面料能做什么，或面料允许你做什么。这种方法称为“立体裁剪”或“在台子上工作”，它是时尚设计的重要部分。立体裁剪式的时尚设计，就是将面料定位在人台上，逐步发展服装结构。一件服装可以基于设计草图，通过立体裁剪完成；或者时尚设计师可以在服装设计的开始阶段在人台上摆弄面料产生新的设计。设计师或样板师首先采用白棉布等基本面料在人台上进行立体裁剪，很像雕塑家，将面料捋平、弄褶、接缝和别合，直到取得希望的形状，然后在面料上标上铅笔线，打剪口，注释衣片，以免后来弄混淆，再从人台上取下来放平。正是这种立体裁剪得到了第一个样板，然后复制印迹和线条，用尺将线条画圆顺并检查，再进行缝制，就得到第一件样衣，然后在人体上试穿。

专业词汇

1. drape vt. 立体裁剪
2. tape measure 卷尺
3. check vt. 检查
4. fullness n. 宽松度
5. dummy n. 人体模型
6. calico n. 白棉布
7. crease v. 折缝
8. splice vt. 拼接
9. annotation n. 注释
10. flatten vt. 弄平
11. smooth vt. 使平滑

The two processes can both be used side by side, neither is more correct, and there are times when one will be more effective than the other depending on the design and personal reference. At their most basic levels of distinction flat pattern making will be quicker for designs which are closer to existing pattern blocks, for designs which are closer to the body or more graphic in nature or for those pattern makers who feel more comfortable working in this method. For designs which are more organic in nature, have more volume or more flowing use of fabric then drape is going to be an easier choice, especially for those who like to see and experiment with the design straight up as they go along, rather than committing to one pattern straight off.

There are also times when designers will work directly on the stand using real fabric and to be able to apply embellishment or build up texture while the design is holding the shape of the human body. But in most cases the designer will want to be able to reproduce the design and will need to have a flat pattern for production at some stage of the process.

If you've ever tried to work on the stand then you will know that the fabric will start to tell you what it will and will not do. There is a certain sensitivity to working in this way, of allowing the fabric to fold[12] just so, or be caught in such a way.

11.2 Why should fashion designers learn how to drape?

While the majority of companies in the fashion industry no longer use draping as part the design process, draping is a key skill which allows apparel designers to understand what creates a great fit and how to achieve it. If a garment sample fits poorly, a designer who is familiar with how darts and seams give shape to garments can spot what is creating the fit issue and advise the factory how to correct the problem.

However, the art of draping isn't completely lost; in high fashion, conture fashion houses, evening, and lingerie companies most garments are created through draping. When draping a garment, the designer can immediately see what her apparel design will look like on the body, and immediately correct any fit or design problems before putting anything down on paper. In addition, some apparel designs are just impossible to make via flat patternmaking and need to be draped first. And some fabrics need to be experimented with on a dress form to see how they behave.

这两种方法可以一并使用，没有谁对谁错。何时采用何种方法更有效，取决于设计和个人的偏好。它们最基本的区别是，如果设计与已有原型样板比较接近，或比较接近身体，或更加图解化，或样板师觉得使用这种方法更加舒适，那么平面制作样板的方法更快；而设计比较复杂、面料更具立体感或悬垂性，特别是有些人喜欢经过实验，看到设计变成了现实，而不是变成非真实的样板，则选择立体裁剪更加便利。

有些时候，设计师直接使用真实面料在人台上进行立体裁剪和修饰；或者直接在人体设计出有风格特征的款式。但是很多情形下，设计师想在设计的基础上再设计时，就需要在此时有平面纸样。

如果你曾经尝试过在人台上立体裁剪，面料将告诉你它能做什么和不能做什么。这种方式很灵活，可知道面料如何折叠或把控。

11.2 为何时尚设计师要学习立体裁剪？

在时尚行业，很多大公司不再使用立体裁剪作为设计过程的一部分，立体裁剪是一种重要的技能，它使服装设计师知道什么是完美合体的设计和如何取得合体。如果一件服装样品很不合体，而且设计师知道如何使用省道和分割塑造服装的形，那么就能知道在哪个部位解决合身问题，建议工厂如何修改这些问题。

然而，立体裁剪并没有完全失它的艺术性；在高级时尚中，高级时尚屋、晚装和贴身内衣公司，大多数服装都是采用立体裁剪的方法设计的。立体裁剪时，设计师能够立刻看到她设计的服装穿在人体的样式，并纠正不合体的地方，或纠正所有设计方面的问题，然后在将衣片平展在纸上。此外服装设计通过平面裁剪不可能实现，需要首先采用立体裁剪。一些面料需要在人台上进行试验，以了解它们的特性。

专业词汇

12. fold vt. 折叠

While draping for apparel design may seem like a daunting and tedious approach to creating patterns, it's actually one of the more creative parts of the fashion design process. The designer's medium is fabric, which he or she manipulates[13] with skillful hands in the creation of simple to complex designs. Draping has many attributes. Draping allows the designer to evaluate the drape at each step to assure that line and balance are in harmony① with design. A misplaced styleline can be easily changed for greater appeal. The designer can manipulate the fabric, deciding where to place darts, tucks[14] and other design elements and adding fullness for gathers, flares, or diagonal[15] until he or she is visually satisfied. These eliminate the need to guess, as other pattern-making systems may require. There are designs that can be developed using any pattern-making method, but designs require only draping to control the look and esthetics② of the completed garment. The designer also has the advantage of being an eyewitness to and a participant in the evolving shapes, from a lifeless piece of muslin to a perfect replica of the design in three-dimensional form. Creating designs by draping may take a bit longer, but the experienced designer has learned to compete and is willing to combine the skills of draping with other methods to produce garments more quickly. Draping has flexibilities that are not offered by any other system, giving draping the greater advantage. Playing with the way fabric folds and hangs[16] on the body is a fun way to create new fashion designs that you wouldn't have thought of sitting in front of a sketchbook!

11.3 Fabric characteristics and term

Fabrics are a powerful medium in the hands of the designer whose creativity is stimulated by the array of colors, boldness[17] of prints[18], and innovative[19] textures that are offered each season. Knowledge of the characteristics of fabrics and the distinctions among them enables designer to select the most appropriate fabric for the purpose of the garment.

Fabrics are classified according to quality, structure (weave[20], knitted[21], fused[22] nonwoven-fused, or plain[23]), texture, weight, and hand (how fabric feels to the touch[24]). They are contrasted by being either crisp[25] or soft[26], thick[27] or thin[28], heavyweight or lightweight, loosely or firmly woven, flat or textured, silky or rough, transparent[29] or opaque[30], and sleazy[31] or luxurious. To evaluate the quality of a fabric before purchasing, the designer should touch and feel the fabric, observe how well it drapes by holding and raising one end to allow the folds to fall naturally, and reaction of the fibers after being crushed[32] in the hand. There is much to learn about fabrics. In addition to reading books devoted to textiles, you should collect swatches when buying fabrics and catalogue them by width and content. In this way, you can create a personal reference library.

用立体裁剪将服装设计转变成纸样，这种方法似乎令人却步和乏味，但确实是时尚设计过程中最具创造性的一部分。设计师的媒介是面料，他或她用一双灵巧的手处理从简单到复杂的设计。立体裁剪有很多特征，使设计师能够在每一步骤评估面料的状况，保证线和平衡与设计取得和谐，错误的样式线条能轻易地改变服装的整体形象。设计师能够操作面料，决定在哪里放置省道、折叠和其他的设计元素，用抽褶、展开或斜向面料增加丰满度，直到他或她满意为止，而不用像其他裁剪方法那样需要猜测。有很多设计可以任意采用某种纸样裁剪的制作方法，但是设计需要通过立体裁剪控制成品服装的样式和美感，有利于设计师目睹和参与形态的演变过程，从毫无生气的白坯布到完美的三维设计形态。通过立体裁方法进行的设计可能需要的时间长一点，但有经验的设计师有能力和乐意结合其他方法，使立体裁剪更加快速。立体裁剪具有其他任何裁剪方法没有的灵活性，这就是立体裁剪的优势。将面料在人体上折叠和悬挂就能创造崭新的时尚设计，而不必坐下来面对速写本，真是一件很有趣的事情。

11.3 面料特征和术语

面料是设计师进行设计的有效媒介，每一季提供的各种色彩、大胆印花和创新肌理的面料，都能激起设计师的创意。熟悉面料的特征和它们之间的差别能使设计师选择最适合服装用途的面料。

面料根据品质、结构（梭织、针织、热熔、非织造热熔或平纹），肌理、重量和手感（手的触感）分类。它们的对比是：脆硬或柔软、厚或薄、重或轻，松织或紧纺、绸感或糙感、透明或不透明，便宜或昂贵。在购买前设计师要评估面料的品质，触摸和感受面料；将面料折叠起来，然后拎起面料一角，松开面料让它自然下垂了解它的悬垂性；将面料在手中挤压，观察纤维的反应，有很多关于面料的知识需要学习。除了阅读面料书籍，在你购买面料时，应该收集面料小样，并且根据它们的应用范围和成分进行分类。通过这种方法，你可以创造一个个人资料图书馆。

专业词汇

13. manipulate vt. 操作
14. tuck n. 箱型褶裥
15. diagonal n. 对角线
16. hang vt. 悬挂
17. boldness adj. 大胆
18. print n. 印花
19. Innovative adj. 创新的
20. weave n. 织物
21. knit n. 针织
22. fuse vi. 热熔
23. plain n. 平纹
24. Touch n. 手感
25. crisp adj. 挺括
26. soft adj. 柔软
27. thick adj. 厚的
28. thin adj. 薄的
29. transparent adj. 透明的
30. opaque adj. 不透明
31. sleazy adj. 质地薄的
32. crush vt. 挤压

通用词汇

① harmony n. 协调
② esthetic adj. 审美的

11.3.1 Muslin

Fashion draping and fitting are usually done with muslin to resolve any design and fitting issues of a garment before cutting the pattern in real fabric. In draping the drapery of the chosen fabric is very important because it affects the finished look of garment. The muslin should be chosen carefully according to the real fabric used in design. Garments made of woven[33] fabrics should be draped with muslin or cheaper fabric where weft[34] and warp[35] yarns can be easily seen. The quality and the hand of the muslin should represent the fabric that is used in design. Garments made of knit should be draped by using inexpensive knitted fabric which has the same stretch[36] with the real fabric used in design.

Muslin fabric, the most common material used for draping, is inexpensive and falls loosely over the dress form, making it easy to manipulate to create different looks. It is a plain-woven fabric made from bleached[37] or unbleached yarns in a variety of weights, including:

(1) Lightweight muslin: It represents the drapery of natural and synthetic silk, light cotton fabrics and lingerie fabrics.

(2) Medium weight Muslin: It simulates the wool and medium weight cottons.

(3) Course muslin: It simulates heavy weight wool and cotton fabrics.

(4) Canvas muslin: It is used to drape denim, fur and some heavy weight fabrics.

11.3.2 Grains and lines

Selvage[38]: The narrow, firmly woven finished edge[39] on both sides of the fabric length. To release the tension[40], clip[41] along the selvage edge.

Grain: The direction in which a fiber is woven or kintted.

The weft and warp directions or the grain line refers to the orientation of yarns in woven fabrics. The weft and warp yarns are perpendicular[42] to each other in the weaving loom[43].

The warp yarns form the lengthwise grain and parallel[44] with the selvage. It is twisted[45] more tightly than the crosswise grain and also called straight grain. The weft yarns form the crosswise grain woven across the fabric from selvage to selvage.

The drape of the fabric varies in different directions since the warp yarns resist[47] stretching most of the garments are cut in lengthwise[48] direction, perpendicular to the hem[49]. As the weft yarns stretch more than warp yarns, they can easily adapt the movements of the body. So the crosswise grain of the fabric is mostly used around the body. The crosswise grain of the fabric is very seldom placed vertically because the yarns will relax[50] and the garment will droop.

Any angle that falls between crosswise grain and length wise grain is called the bias. True bias[51] is at a 45-degree angle to lengthwise and crosswise grain. True bias has maximum give and stretch, easily confirming to the contour of the figure. Flares and cowls drape best when on true bias. The designers use bias cut to give the dresses more shape and fullness. Because of the stretch and distortion[52] of the bias, the cutting and sewing of bias cut fabric patterns need more attention.

11.3.1 平纹细布

时尚立体裁剪通常使用平纹细布解决服装上任何设计和合身的问题，然后再在真实的面料上裁剪衣片。在立体裁剪中，选择立体裁剪用的面料很重要，它影响服装成品的面貌。要根据设计中使用的真实面料仔细挑选平纹细布。梭织面料服装，应该使用平纹细布或便宜面料立体裁剪，纬向和经向要很清晰。平纹细布的质量和手感必须与设计中使用的面料相似。针织面料服装的立体裁剪要使用便宜的针织面料，要和设计中的真实面料有相同的弹力。

平纹细布是立体裁剪时普遍采用的面料，便宜而且能随意地覆盖在人台上，容易操作，创建不同的外观。它是一种平纹纺织面料，用漂白或不漂白纱线制造，重量有多种，包括：

（1）轻型平纹细布：替代天然和人造丝绸、轻型棉织物和内衣面料的立体裁剪。

（2）适中型平纹细布：替代羊毛和重量适中的棉质面料。

（3）粗糙平纹细布：替代重型羊毛和棉织物。

（4）帆布平纹细布：替代牛仔、裘皮和一些重型面料。

11.3.2 纹理和纹路

布边：面料长度方向两边窄的、坚固的织边。为了释放张力，沿着布边将它剪去。

纹理：纺织或针织时纱线的方向。

纬向和经向或纹理是指面料纺织时纱线方向。纬向和经向在织布机上相互垂直。

经纱构成纵向的纹理，平行于布边，它的绞捻度比横向纹理紧，也称为直向纹理。纬纱构成了横向纹理，从布边的一边到另一边。

立体裁剪的面料可以变化不同方向，因为经纱有抵抗拉伸的力量，大多数服装裁剪时都采用经纱与底边垂直。因为纬纱比经纱容易拉伸，它们能够适合身体的运动，因此，面料的纬纱大多用来包围身体。面料纬向很少垂直放置，因为纱线会松弛，服装也将会下垂。

在纵向和横向之间的任何角度都称为斜向。正斜是指与纵向和横向纹理都成45°角度。正斜具有最大的弹性和拉伸力，可有效地体现身体的轮廓线。喇叭状和垂荡形最好采用正斜。设计师使用正斜裁剪，使连衣裙更具有形和丰满性。因为斜向容易拉伸和变形，裁剪和缝纫时需要特别注意。

专业词汇

33. woven adj. 梭织
34. weft n. 纬纱
35. warp n. 经纱
36. stretch n. 拉伸
37. bleach vt. 漂白
38. selvage n. 布边
39. finished edge 实边
40. tension n. 张力
41. clip vi. 剪掉
42. perpendicular adj. 垂直的
43. loom n. 织布机
44 parallel adj. 平行的
45. twist vt. 绞捻
46. crosswise adj. 横的
47. resist v. 抵抗
48. lengthwise adv. 纵长地
49. hem n. 底边
50. relax vt. 放松
51. true bias 正斜
52. distortion n. 变形

11.4 Model form

For the past 140 years, forms have adapted to the whims of fashion by constantly modifying in shape and measurement to satisfy the needs of changing silhouettes. Original forms were shapeless, willow-caned models with woven mounds that were padded to individual specifications. Today's forms are partially made by hand. They are framed in metal, molded with papier-mâché, laid over with canvas, and covered in a princess garment of linen[53]. The seam lines of the cover garment set the boundaries between the front and back bodice and skirt. The waistline[54] seam defines the upper and lower torso.

A new innovation in form making creates forms that feel flesh[55] to the touch and can be penetrated with pins without harm to the form. Forms of today represent the most common dimensions within each size group of males and females, children to adults. Forms come with detachable arms and legs, and collapsible③ shoulders for ease of entry.

11.5 Basic dress foundation

Fashion designers drape garments in sections i. e. : front bodice, back bodicc, fronl skirt, back skirt etc and only the right side of the garment is draped, unless the apparel design is asymmetrical[56].

The simplicity of the basic dress makes it an ideal choice for the introduction to draping. The principles of draping and related draping techniques will be applied to complete the drape of the basic dress. After mastering the fundamentals of the draping process, the designer will advance with confidence to more complicated designs that build on these applications.

The dress is draped to fit the dimensions of the form, or figure and bridges hollow④ areas between the bust, buttocks[57], and shoulder blades[58]. Ease is added for comfortable movement without the appearance of stress. The sleeves hang with relaxed arms in the perfect alignment⑤. The skirt straight from the widest part of the hips and the hem is parallel to the floor. A number of darts control the fit of the garment by taking up remaining excess where it is not wanted, such as the ends of the radiating[59] points of the bust, buttocks and shoulder blades.

The foundation of the basic dress is related to all draped garments by the application of the draping and technical principles learned in its creation, including the relationships among line, balance, and fit. To drape a perfect garment takes time and patience. Every accomplished designer knows that hard work and perseverance⑥ is the key to perfection.

The following information discusses three draping principles and techniques that apply to the process of developing designs.

11.4 人体模型

在过去140年里，人台不断修改形状和尺寸，以适合时尚的奇思怪想，满足廓型不断变化的需求。最初人台是用不定型的柳条编织成模型，再根据个体的尺寸进行衬垫。今天有些人台是手工制作，其框架采用金属，模型用纸浆，再用帆布包裹，上面再覆盖一层有公主线分割的亚麻布。上面一层的服装有缝线，区分上身前后片和裙子的前后片。腰线界定了躯干的上部分和下部分。

一种新制造发明的人体模型是触摸时更具肉感，可以插针而不会对模型造成损害。今天，有男性、女性，孩子、成人不同群体平均尺寸的人体模型。模型有一个可拆卸的手臂、腿和可拆卸的肩，是为了穿衣方便。

11.5 基础服装知识

服装设计师按照服装的不同部位进行立体裁剪，例如前衣身、后衣身、前裙片、后裙片，等等，通常只立体裁剪服装的右半部分，如果服装设计是不对称的，则立体裁剪全部。

简单的基础服装，是介绍立体裁剪的理想选择。基础服装的立体裁剪中应用了立体裁剪的基本原理和相关立体裁剪技术。掌握了立体裁剪基础知识以后，设计师将更有自信地把基础知识应用到复杂的设计中。

基本立体裁剪就是贴合人体模型或人体的尺寸，弥合胸部、臀部和肩胛骨区域的空隙。增加松量是为了运动的舒适性，服装不会出现拉扯。在手臂放松时袖子处于精确校准的状态。裙子从最宽臀部部位到裙摆要直，裙摆平行于地板。一些褶裥能控制服装的贴合性，即通过收掉不需要的多余量，例如以胸部、臀部和肩胛骨等部位为末端点的放射状省道。

基础服装的立体裁剪知识可以应用到所有服装的立体裁剪中，包括线条、平衡和合身的关系。立体裁剪使服装达到完美状态需要时间和耐心。每一个有成就的设计师都知道，艰苦工作和持之以恒是获得完美服装的关键。

以下信息讨论在设计进展过程中立体裁剪应用到的三个基本原则和技术。

专业词汇

53. linen n. 亚麻布
54. waistline n. 腰围线
55. flesh n. 肉体
56. asymmetrical adj. 不对称的
57. buttock n. 臀部
58. shoulder blade 肩胛骨
59. radiate vt. 放射

通用词汇

③ collapsible adj. 可折叠的
④ hollow adj. 空的
⑤ alignment n. 队列
⑥ perseverance n. 坚持不懈

11. 5. 1 Three draping principles

Principles of the dart excess—The dart is a fitting device[60] that controls remaining excess within the boundaries of the marked drape. The dart is identified by its wedge[61] shape, formed in the process of draping. The width depends on the amount of excess; length depends on distance from its source.

Principles of the cross-grain—The cross-grain placed on the form parallel with the floor will balance a garment and divide fullness above and below a guideline. Fall of the cross-grain creates bias and flare or lift for drapery, pleats, and gathers.

Principles of a balanced sleeve—A well-balanced sleeve aligns with or is slightly forward of the side seam of the form. A sleeve will hang out of alignment if the shoulder line and/or side seam of the form have not been adjusted. The sleeve must align with the angle of the clients' arm and stance[62].

11. 5. 2 Three draping techniques

Moving dart excess-excess is moved along the seam lines (boundaries) of the form while smoothing muslin around the bust, or any mound, to locations directed by the design. Excess may be in the form of a dart equivalent in creating the design.

Adding fullness—To create fullness, slash and drop the cross-grain; lift the grain for fullness to create drapery, gathers, and pleats. Use the ratio[63] rule to determine fullness needed. Do this when the dart excess is insufficient.

Contouring—To reveal the contours of the form or figure, remove the bust bridge to allow fabric to be draped into the hollows around the bust (padded for emphasis, or unpadded). Side ease is removed for strapless gowns[64].

11. 6 Manipulation of draping in couture design

In couture design, draping method is used very effectively. For using draping, designer should know the human anatomy[65], fibers and fabric. The silhouette and the fabric should be in harmony since the weight, texture, hand and drape of the fabric might not be appropriate for some silhouettes. If the designer wants to create sculpted look, crispy woven fabric should be chosen. If he works with very soft fabric, the design will follow the curves of the boy and drape gently.

In a couture house after the appropriate fabrics are selected, the designer works with the fabric to see how it hangs on the lengthwise grain, crosswise grain or bias. Then the designer starts sketching for his collection and sketches are distributed to the workrooms of the house. More structured garments like jackets and coats are sent to tailoring room, night gowns, dresses and skirts which need to be draped are sent to dressmaking room. Dressmaking room is where designs are draped and sewn on a dress form temporarily. In the dress making room, muslin which has similar drape and hand with the real fabric that is going to be used in dress is chosen. Dress forms are padded according to body measurements of the individual.

11.5.1 三个立体裁剪基本原则

省道余量原则——省道是一种合身设计，它能控制立裁中边界标记以内多余的量。省道形状为楔形，在立体裁剪过程中形成。其宽度依赖于多余的量，长度根据它到源头的距离。

横向原则——横纱向放在人台上时要与地面平行，使服装取得平衡，在引导线上方和下方作丰满度的分配。横纱向下降产生斜向和展开，横纱向提升产生垂荡、褶裥和抽褶。

袖子平衡原则——在袖子取得很好的平衡状态下，与人台侧缝一致或略微向前。如果肩线和侧缝不修正，袖子可能歪着。袖子必须与客户手臂的角度和姿态一致。

11.5.2 三种立体裁剪技术

转移余量——余量沿着人台缝线（边界线）移动到设计指向的部位，但是，围绕胸部或任何隆起部位的布要捋平。

增加丰满度——为了取得丰满状态，剪切和下降横纱向，或提升横纱向产生悬垂、抽褶和褶裥。当省道多余量不够时，使用比率规则满足丰满度的需求。

轮廓线——为了显示人台或人体曲线，将胸部两乳之间搭建的布条去掉，使面料覆盖到胸部周围凹陷的部位（此时为了强调胸部采用衬垫，或者胸部不衬垫），无带裙装去掉侧缝的放松量。

11.6 高级时尚设计中立体裁剪的运用

在高级时尚设计中，使用立体裁剪方法十分有效。为了使用立体裁剪，高级时尚设计师必须知道人体解剖、纤维和面料的知识。廓型和面料必须协调，因为面料的重量、肌理、手感和悬垂性可能不适合一些廓型。如果设计师想创造雕塑般的形象，应该选择硬挺的面料。如果他采用十分柔软的面料，设计将跟随身体曲线微微下垂。

在高级时尚屋，在选择了适当的面料以后，设计师就要开始审视面料，查看面料悬挂时它的直纱向、横纱向和斜纱向如何变化。然后，设计师开始为他的设计画草图，再将草图送到各个工作车间。夹克和外套这些多结构的服装要被送到精致裁剪的工作车间；需要立体裁剪的晚装、连衣裙和裙子被送到女装裁缝车间。需要立体裁剪的设计和缝制时需要人台的都归入女装裁缝车间。在立体裁剪车间里，选择与设计中使用的真实面料相类似的坯布。人台要根据个人的身体尺寸进行衬垫。

专业词汇

60. device n. 装置
61. wedge n. 楔形
62. stance n. 站姿
63. ratio n. 比率
64. strapless gown 无吊带裙袍
65. anatomy n. 解剖

Then the design chosen is draped on the dress form by a sample maker. By this way basic patterns for the garment are created, this process takes 4 to 8 hours according to the complexity of the design. These patterns are basted together for the first fit. Even though patterns are working patterns, every detail like buttonholes[66], trims are carefully applied. After the fit, the patterns are corrected; this corrections and fittings are repeated until the designer is satisfied with the look.

Sometimes while draping, the designer doesn't use any inner structure and the garment shows the lines of the body. Madeleine·Vionnet' bias cut dresses and Madame Gres' dresses are the best example for this type of dresses. Sometimes designers use inner structure to give shape, and this structure is like a skeleton that holds the drapes of the garment in place. We can see this type of design philosophy in Charles James garments.

The visual and functional properties of a garment are directly connected to the relation between body and the garment. A garment might have a good design but might not give enough freedom for movements of body. For this reason, the two dimensional patterns should follow the three dimensional form of the body and the best way of providing the perfect fit is draping.

There are several advantages of using draping in garment design. First of all since the three dimensional effect of the garments is very different than the effect on paper; a new and much innovative design can be created by draping. Since the designer works with the fabric, he can use the fabric with its maximum potential in his designs. Sometimes the drapes and folds of the fabric may be very different than it is sketched on paper.

By the help of draping a wrong fitted garment is quickly recognized and can be corrected. When the body measurements are different than the standard measurements of the flat pattern method, the patterns may not fit to the body. Since draping is done on a model that is modified according to the body measurements, the patterns will perfectly fit to the body. Draping is a procedure that affects the garment quality directly. Although draping is very expensive and time consuming when it is practiced correctly it gives the best result.

然后，挑选出来的设计由样衣师在人台立体裁剪。通过这种方法，得到服装的基本衣片，根据设计的难易程度，这个过程大约需要4~8个小时。这些衣片被粗缝在一起，进行第一次试穿。即使样板是工作样板，每个细节、装饰都要仔细地加上，例如纽扣洞。在试穿之后，衣片要修改，这种修改和试穿要进行好多次，直到设计师满意为止。

有时设计师使用立体裁剪时不使用任何内部结构，服装就能显示身体的线条。马德琳·维奥内（Madeleine Vionnet）的斜裁连衣裙和格瑞斯夫人（Madame Gres）设计的连衣裙是这种服装的最好事例。有时设计师使用内部结构塑造形状，这种结构就像骨骼，使服装处于正确的平衡状态。我们可以在查尔斯·詹姆斯（Charles James）的服装设计中看到这种结构原理。

一件服装的视觉和功能特征直接与身体和服装之间关系相关。也许服装设计得很好，但是没有足够空间可以自由活动。正是这个原因，二维纸样应该遵循三维身体模型，提供完美合身的最好方法是立体裁剪。

在服装设计中，使用立体裁剪有很多优势，首当其冲的是服装的三维效果与纸上的效果明显不同，一种新的、十分有创意的设计可以通过立体裁剪实现。因为设计师的工作对象是面料，他可以在设计中发挥面料最大的潜在特征。有时，面料的悬垂和折叠与纸上的草图完全不同。

立体裁剪有助于很快识别服装哪里不合身，并做出快速的修改。当身体尺寸与平面纸样的标准尺寸不同时，样板可能不合身。而立体裁剪是在人体上实现的，可以根据身体的尺寸做修改，样板将能很好地适合身体。立体裁剪是直接影响服装质量的一种工序，尽管立体裁剪的花费贵了一些，在不断修改时消耗时间，但是它能够取得最好的效果。

专业词汇

66. buttonhole n. 扣眼

Exercises

(1) Understanding the text.

Read the text and answer the following questions.

1) What are the advantages of draping?

2) Since draping has become the mainstream of garment design and plate making, how can the draping be promoted in China?

3) How can you improve your draping ability?

(2) Building your language.

The following words and expressions can be used to talk about draping thoughts. Choose the right ones to fill in the blanks in the following sentences. Change the form where necessary.

commission	culminate	essential	garments
mannequin	dimension	advances	volume

Draping is the art of using cotton muslin to create a fashion design directly on a ________. It is an ________ skill for fashion designers. In this book, Karolyn Kiisel presents a series of step-by-step projects, creating real ________ in classic styles, that are designed to develop skills from the most basic to more advanced techniques.

Starting with the basics of choosing and preparing the dress form for draping, the article ________ through pinning, trimming, and clipping, and creating shape using darts and tucks, to adding ________ using pleats and gathers, and handling complex curves. Advanced skills include how to use support elements such as shoulder pads, under layers, and petticoats, and how to handle bias draping. The article ________ with a chapter on improvisational skills.

Each skill and technique throughout the article is explained through specially ________ step-by-step photographs and line drawings that bring the art of creating women swear in three ________ to life.

(3) Sharing your ideas.

After learning about draping in this section, are you eager to share your knowledge? Please write a short introduction (of around 300 words) about draping. Try to make full use of what you've learned from Text, including the relevant information from the reading text as well as words and expressions.

Keys to Exercises

Unit 1

Building your language

1. a variety of
2. ranging from
3. immersive
4. interface
5. reveal
6. redefine
7. tool
8. seeking

Unit 2

Building your language

1. durable
2. outlast
3. sustainability
4. measured
5. notion
6. sturdily
7. counterparts
8. worst

Unit 3

Building your language

1. calming
2. survey
3. participants
4. stimulation
5. psychology
6. relaxation
7. associated
8. luxury

Unit 4

Building your language

1. 杰奎琳的打扮有时简洁到只穿上下两件式、珠宝装饰领口的制服式套装，外加一顶礼帽。
2. 推档就是放大或缩小标准生产纸样的尺寸以做出一个完整的尺寸范围的方法。
3. 联合会由三个部门组成：高级时装部、女装成衣部和男装部。
4. 英国的经典时装在世界上享有很高的声誉。
5. 1911 年美国销售的服饰有一半是靠促销（打折、降价促销）卖出的。

Unit 5

Building your language

1. available
2. comfort
3. style
4. ensure
5. categories
6. outfit
7. ocassion
8. fashion

Unit 6

Building your language

1. 为迎合流行，腰带可以用各种材料制作，如网，塑料，穗带，锁链，绳子甚至橡胶。
2. 鞋类公司的产品种类繁多，价格各有不同。
3. 在通晓市场营销策略，制作吸引美国消费者的畅销时装方面，美国设计师做的非常优秀。
4. 加拿大和墨西哥是美国最大的贸易伙伴，当然，美国与这两国的贸易也占据了它们贸易总额的 2/3。
5. 社交礼服包含一些适合特殊场合穿着的服装，如或长或短的晚礼服、时髦的长裤套装、晚礼服和新娘礼服等。

Unit 7

Building your language

1. break into	2. perseverance	3. passion	4. exclusive
5. mental	6. discerning	7. designers	8. rent

Unit 8

Building your language

1. original	2. glamorous	3. elegance	4. leading
5. ready-to-wear	6. accessible	7. fabric	8. incorporated into

Unit 9

Building your language

1. 消费者对于安全无污染产品的喜好，使得一些厂商寻找天然的有机面料。
2. 在通晓市场营销策略，制作吸引美国消费者的畅销时装方面，美国设计师做得非常优秀。
3. 由于科技的进步和人造合成材料的广泛使用，全美国出售的纽扣中只有不到 10% 是由天然物质制作的。
4. 鞋类公司的产品种类繁多，价格各有不同。
5. 这种处理方法使这种布料永久保持形状和尺寸，增强布料的弹性。

Unit 10

Building your language

1. anniversary
2. promotion
3. integrate
4. combing with
5. fibers
6. enterprises
7. textile
8. consumption

Unit 11

Building your language

1. mannequin
2. essential
3. garments
4. advance
5. culminates
6. volume
7. commissioned
8. dimensions

References

[1] Sue Jenkyn Jones. *Fashion design* [M]. Second Edition. London: Laurence King Publishing, 2005.

[2] Uche Okonkwo. *Luxury Fashion Branding: Trends, Tactics, Techniques* [M]. Berlin: Springer, 2007.

[3] Marcarena San Martin. *Field Guide: How to be a fashion designer* [M]. New York: Rockport Publishers, 2009.

[4] Lisa J. Springer. *Becoming a fashion designer* [M]. New York: The Simon & Schuster Speakers Bureau, 2014.

[5] Richard Sorger, Jenny Udale. *The Fundamentals of Fashion Designing* [M]. West Sussex: AVA Publishing SA, 2006.

[6] Evelynl Brannon. *Fashion Forecasting* [M]. Chicago: Fairchild publications Inc, 2000.

[7] Anette Fischer. *Basics Fashion Design Construction* [M]. New York: Thames & Hudson, 2008.

[8] Helen Joseph-Armstrong. *Draping for Apparel Design* [M]. Chicago: Fairchild Publications, 2013.

[9] Lise Skov, Else Skjold, Brian Moeran, Frederic Larsen and Fabian F. Csaba. *The Fashion Show as an Art Form* [J]. Creative Encounters Working Paper, 2009, (1): 1-37.

[10] Colin Gale, Jasbir Kaur. *Fashion and Textiles* [M]. Oxford: Berg Publishers, 2004.

[11] Elina Timone. *Adapting Design to Foreign Markets— A Case Study of Three Finnish Fashion Firms* [J]. Academic journal of Aalto University, 2012 , 13114 (1): 118-129.

[12] Mersiha Memic, Frida N. Minhas. *The fast fashion phenomenon Luxury fashion brands responding to fast fashion* [J]. University of Borås/swedish School of Textiles, 2011, 23209 (2): 41-52.

[13] Kasra Ferdows, Michael A. Lewis, Jose A. D. Machuca. *Zara's Secret for Fast Fashion* [M]. Saarbrücken: VDM Verlag Dr. Müller, 2005.